METHODS IN MOLECULAR BIOLOGY

Series Editor
John M. Walker
School of Life and Medical Sciences
University of Hertfordshire
Hatfield, Hertfordshire, UK

More information about this series at:
http://www.springer.com/series/7651

Bioreactors in Stem Cell Biology

Methods and Protocols

Edited by

Kursad Turksen

Ottawa Hospital Research Institute
Ottawa, ON, Canada

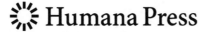 Humana Press

Editor
Kursad Turksen
Ottawa Hospital Research Institute
Ottawa, ON, Canada

ISSN 1064-3745   ISSN 1940-6029   (electronic)
Methods in Molecular Biology
ISBN 978-1-4939-8213-4   ISBN 978-1-4939-6478-9   (eBook)
DOI 10.1007/978-1-4939-6478-9

Printed on acid-free paper

This Humana Press imprint is published by Springer Nature
The registered company is Springer Science+Business Media LLC New York

Preface

Our understanding of stem cells and their potential to repair and regenerate tissues and organs has been increasing steadily over the last couple of decades. However, for many applications, we now recognize that we need to translate what has been learned primarily in tissue culture dishes to approaches supporting scale-up not only to large quantities of cells but also to heterogeneous cell constructs. Notable advances are being made in these latter approaches, prompting us to collect a variety of representative protocols that may facilitate important modifications and novel approaches by both established and new investigators in this area. With this objective in mind, I would like to thank all the contributors for their contributions to this volume.

I also thank Dr. John Walker, Editor in Chief of the *Methods in Molecular Biology* series, for giving me the opportunity to put together this volume.

I am grateful to Patrick Marton, Senior Editor of the *Methods in Molecular Biology* series, for his continuous support throughout the project.

A special "thank you" goes to David Casey, Editor of the *Methods in Molecular Biology* series, for his tireless efforts in supporting me to eliminate any missing items and helping to finalize the volume.

Ottawa, ON, Canada   *Kursad Turksen*

Contents

The original version of this book was revised.
An erratum to this book can be found at DOI 10.1007/978-1-4939-6478-9_5001

Contributors

SAEED ABBASALIZADEH • *Department of Stem Cells and Developmental Biology, Cell Science Research Center, Royan Institute for Stem Cell Biology and Technology, ACECR, Tehran, Iran*

SPIROS N. AGATHOS • *School of Life Sciences and Biotechnology, Yachay Tech University, San Miguel de Urcuquí, Ecuador; Earth and Life Science Institute - Laboratory of Bioengineering, Catholic University of Louvain, Louvain-la-Neuve, Belgium*

PAYAM AKHYARI • *Research Group for Experimental Surgery, Department of Cardiovascular Surgery, Medical Faculty, Heinrich Heine University Dusseldorf, Dusseldorf, Germany*

WALAA ALMUTAWAA • *Department of Biochemistry and Molecular Biology, University of Calgary, Calgary, AB, Canada*

PAULA M. ALVES • *iBET, Instituto de Biologia Experimental e Tecnológica, Oeiras, Portugal; Instituto de Tecnologia Química e Biológica António Xavier, Universidade Nova de Lisboa, Oeiras, Portugal*

FRANCISCA AREZ • *iBET, Instituto de Biologia Experimental e Tecnológica, Oeiras, Portugal; Instituto de Tecnologia Química e Biológica António Xavier, Universidade Nova de Lisboa, Oeiras, Portugal*

PREETI ASHOK • *Department of Chemical and Biological Engineering, Tufts University, Medford, MA, USA*

HUG AUBIN • *Research Group for Experimental Surgery, Department of Cardiovascular Surgery, Medical Faculty, Heinrich Heine University Dusseldorf, Dusseldorf, Germany*

JIWOO BAE • *CHA Stem Cell Institute, Department of Biomedical Science, CHA University, Gyeonggi-do, Republic of Korea*

HOSSEIN BAHARVAND • *Department of Stem Cells and Developmental Biology, Cell Science Research Center, Royan Institute for Stem Cell Biology and Technology, ACECR, Tehran, Iran; Department of Developmental Biology, University of Science and Culture, ACECR, Tehran, Iran*

JENNIFER G. BARRETT • *Department of Large Animal Clinical Sciences, Marion duPont Scott Equine Medical Center, Virginia Tech, Leesburg, VA, USA*

GUNNAR BERGSTRÖM • *Division of Biotechnology, Department of Physics, Chemistry and Biology (IFM), Linköping University, Linköping, Sweden*

JEFF BIERNASKIE • *Department of Comparative Biology and Experimental Medicine, Faculty of Veterinary Medicine, Alberta Children's Hospital Research Institute, Hotchkiss Brain Institute, Cumming School of Medicine, University of Calgary, Calgary, AB, Canada; Faculty of Veterinary Medicine, University of Calgary, Calgary, AB, Canada*

CATARINA BRITO • *iBET, Instituto de Biologia Experimental e Tecnológica, Oeiras, Portugal; Instituto de Tecnologia Química e Biológica António Xavier, Universidade Nova de Lisboa, Oeiras, Portugal*

ANTHONY CALLANAN • *Department of Mechanical and Aeronautical Engineering and MSSI, Centre for Applied Biomedical Engineering Research, University of Limerick, Limerick, Ireland; Institute for Bioengineering, School of Engineering, The University of Edinburgh, Edinburgh, UK*

WANKYU CHOI • *CHA Stem Cell Institute, Department of Biomedical Science, CHA University, Gyeonggi-do, Republic of Korea*

JONAS CHRISTOFFERSSON • *Division of Biotechnology, Department of Physics, Chemistry and Biology (IFM), Linköping University, Linköping, Sweden*

NIALL F. DAVIS • *Department of Mechanical and Aeronautical Engineering and MSSI, Centre for Applied Biomedical Engineering Research, University of Limerick, Limerick, Ireland*

ZHILI DENG • *State Key Laboratory of Stem Cell and Reproductive Biology, Institute of Zoology, Chinese Academy of Sciences, Beijing, China*

ANDRAS DINNYES • *BioTalentum Ltd., Gödöllö, Hungary; Molecular Animal Biotechnology Laboratory, Szent Istvan University, Gödöllö, Hungary*

INA DOBRINSKI • *Department of Comparative Biology and Experimental Medicine, Faculty of Veterinary Medicine, Cumming School of Medicine, University of Calgary, Calgary, AB, Canada; Department of Biochemistry and Molecular Biology, Cumming School of Medicine, University of Calgary, Calgary, AB, Canada*

CAMILA DORES • *Department of Comparative Biology and Experimental Medicine, Faculty of Veterinary Medicine, Cumming School of Medicine, University of Calgary, Calgary, AB, Canada*

ENKUI DUAN • *State Key Laboratory of Stem Cell and Reproductive Biology, Institute of Zoology, Chinese Academy of Sciences, Beijing, China*

JAN-CHRISTOPH EDELMANN • *Imperial College London, London, UK*

YONGJIA FAN • *Department of Chemical and Biological Engineering, Tufts University, Medford, MA, USA*

JOAO N. FERREIRA • *Department of Oral and Maxillofacial Surgery, Faculty of Dentistry, National University of Singapore, Singapore, Singapore*

JÖRG C. GERLACH • *McGowan Institute for Regenerative Medicine, University of Pittsburgh, Pittsburgh, PA, USA*

BAEK SOO HAN • *Research Center for Integrated Cellulomics, KRIBB, Daejeon, Republic of Korea*

JÖRN HÜLSMANN • *Research Group for Experimental Surgery, Department of Cardiovascular Surgery, Medical Faculty, Heinrich Heine University Dusseldorf, Dusseldorf, Germany*

ALEXANDER JENKE • *Research Group for Experimental Surgery, Department of Cardiovascular Surgery, Medical Faculty, Heinrich Heine University Dusseldorf, Dusseldorf, Germany*

NOO LI JEON • *School of Mechanical and Aerospace Engineering, Seoul National University, Seoul, Republic of Korea*

LIZZIE JONES • *Imperial College London, London, UK*

MICHAEL S. KALLOS • *Pharmaceutical Production Research Facility (PPRF), University of Calgary, Calgary, AB, Canada; Biomedical Engineering Graduate Program, University of Calgary, Calgary, AB, Canada; Chemical and Petroleum Engineering, Schulich School of Engineering, University of Calgary, Calgary, AB, Canada*

HENNING KEMPF • *Leibniz Research Laboratories for Biotechnology and Artificial Organs (LEBAO), Hannover Medical School, Hannover, Germany*

JUNG JAE KO • *CHA Stem Cell Institute, Department of Biomedical Science, CHA University, Gyeonggi-do, Republic of Korea*

PETER KOHL • *Institute for Experimental Cardiovascular Medicine, University Heart Centre Freiburg - Bad Krozingen and Medical School of the Albert Ludwigs University Freiburg, Freiburg, Germany; Imperial College London, London, UK*

NAYEON LEE • *CHA Stem Cell Institute, Department of Biomedical Science, CHA University, Gyeonggi-do, Republic of Korea*

SUJI LEE • *CHA Stem Cell Institute, Department of Biomedical Science, CHA University, Gyeonggi-do, Republic of Korea*

SANG CHUL LEE • *Research Center for Integrated Cellulomics, KRIBB, Daejeon, Republic of Korea*

XIAOHUA LEI • *State Key Laboratory of Stem Cell and Reproductive Biology, Institute of Zoology, Chinese Academy of Sciences, Beijing, China*

YAN LI • *Department of Chemical and Biomedical Engineering, FAMU-FSU College of Engineering, Florida State University, Tallahassee, FL, USA*

ARTUR LICHTENBERG • *Research Group for Experimental Surgery, Department of Cardiovascular Surgery, Medical Faculty, Heinrich Heine University Dusseldorf, Dusseldorf, Germany*

LIANG LU • *Medical College of Georgia at Augusta University, Augusta, GA, USA*

TENG MA • *Department of Chemical and Biomedical Engineering, FAMU-FSU College of Engineering, Florida State University, Tallahassee, FL, USA*

TENG MA • *Department of Chemical and Biomedical Engineering, Florida State University, Tallahassee, FL, USA*

CARL-FREDRIK MANDENIUS • *Division of Biotechnology, Department of Physics, Chemistry and Biology (IFM), Linköping University, Linköping, Sweden*

ALESSANDRO MARTURANO-KRUIK • *Department of Chemistry, Materials and Chemical Engineering "G Natta", Politecnico di Milano, Milano, Italy; Department of Biomedical Engineering, Columbia University, New York, NY, USA*

RAJIV MIDHA • *Department of Comparative Biology and Experimental Medicine, Faculty of Veterinary Medicine, Alberta Children's Hospital Research Institute, Hotchkiss Brain Institute, Cumming School of Medicine, University of Calgary, Calgary, AB, Canada*

SHIVA NEMATI • *Department of Stem Cells and Developmental Biology, Cell Science Research Center, Royan Institute for Stem Cell Biology and Technology, ACECR, Tehran, Iran*

REMI PEYRONNET • *Imperial College London, London, UK; Institute for Experimental Cardiovascular Medicine, University Heart Centre Freiburg - Bad Krozingen and Medical School of the Albert Ludwigs University Freiburg, Freiburg, Germany*

CATARINA PINTO • *iBET, Instituto de Biologia Experimental e Tecnológica, Oeiras, Portugal; Instituto de Tecnologia Química e Biológica António Xavier, Universidade Nova de Lisboa, Oeiras, Portugal*

DERRICK E. RANCOURT • *Department of Biochemistry and Molecular Biology, University of Calgary, Calgary, AB, Canada*

DERRICK E. RANCOURT • *Department of Biochemistry and Molecular Biology, Cumming School of Medicine, University of Calgary, Calgary, AB, Canada; Department of Oncology and Medical Genetics, Cumming School of Medicine, University of Calgary, Calgary, AB, Canada*

URSULA RAVENS • *Institute for Experimental Cardiovascular Medicine, University Heart Centre Freiburg - Bad Krozingen and Medical School of the Albert Ludwigs University Freiburg, Freiburg, Germany; Department of Pharmacology and Toxicology, Medical Faculty "Carl Gustav Carus", Dresden University of Technology (TU Dresden), Dresden, Germany*

LEILI ROHANI • *Department of Biochemistry and Molecular Biology, University of Calgary, Calgary, AB, Canada*

MAHBOUBEH R. ROSTAMI • *Department of Chemical and Biological Engineering, Tufts University, Medford, MA, USA*

SASITORN RUNGARUNLERT • *Department of Preclinical and Applied Animal Science, Faculty of Veterinary Science, Mahidol University, Nakhon Pathom, Thailand*

SADMAN SAKIB • *Department of Comparative Biology and Experimental Medicine, Faculty of Veterinary Medicine, Cumming School of Medicine, University of Calgary, Calgary, AB, Canada; Department of Biochemistry and Molecular Biology, Cumming School of Medicine, University of Calgary, Calgary, AB, Canada*

SÉBASTIEN SART • *Laboratory of Hydrodynamics (LadHyX) - Department of Mechanics, Ecole Polytechnique, CNRS-UMR7646, Palaiseau, France*

EVA SCHMELZER • *McGowan Institute for Regenerative Medicine, University of Pittsburgh, Pittsburgh, USA*

KRISTIN SCHWANKE • *Leibniz Research Laboratories for Biotechnology and Artificial Organs (LEBAO), Hannover Medical School, Hannover, Germany*

DANIEL SIMÃO • *iBET, Instituto de Biologia Experimental e Tecnológica, Oeiras, Portugal; Instituto de Tecnologia Química e Biológica António Xavier, Universidade Nova de Lisboa, Oeiras, Portugal*

LIQING SONG • *Department of Chemical and Biomedical Engineering, FAMU-FSU College of Engineering, Florida State University, Tallahassee, FL, USA*

JIHWAN SONG • *CHA Stem Cell Institute, Department of Biomedical Science, CHA University, Gyeonggi-do, Republic of Korea*

MARCOS F.Q. SOUSA • *iBET, Instituto de Biologia Experimental e Tecnológica, Oeiras, Portugal; Instituto de Tecnologia Química e Biológica António Xavier, Universidade Nova de Lisboa, Oeiras, Portugal*

ANA P. TERASSO • *iBET, Instituto de Biologia Experimental e Tecnológica, Oeiras, Portugal; Instituto de Tecnologia Química e Biológica António Xavier, Universidade Nova de Lisboa, Oeiras, Portugal*

ANG-CHEN TSAI • *Department of Chemical and Biomedical Engineering, FAMU-FSU College of Engineering, Florida State University, Tallahassee, FL, USA*

EMMANUEL S. TZANAKAKIS • *Department of Chemical and Biological Engineering, Tufts University, Medford, MA, USA*

ARANZAZU VILLASANTE • *Department of Biomedical Engineering, Columbia University, New York, NY, USA; Laboratory for Stem Cells and Tissue Engineering, Columbia University, New York, NY, USA*

GORDANA VUNJAK-NOVAKOVIC • *Department of Biomedical Engineering, Columbia University, New York, NY, USA; Laboratory for Stem Cells and Tissue Engineering, Columbia University, New York, NY, USA; Laboratory for Stem Cells and Tissue Engineering, Columbia University, New York, NY, USA*

TYLOR WALSH • *Pharmaceutical Production Research Facility (PPRF), University of Calgary, Calgary, AB, Canada; Biomedical Engineering Graduate Program, University of Calgary, Calgary, AB, Canada*

ALEXANDER WEHRMANN • *Research Group for Experimental Surgery, Department of Cardiovascular Surgery, Medical Faculty, Heinrich Heine University Dusseldorf, Dusseldorf, Germany*

YUANWEI YAN • *Department of Chemical and Biomedical Engineering, FAMU-FSU College of Engineering, Florida State University, Tallahassee, FL, USA*

DANIEL W. YOUNGSTROM • *Department of Large Animal Clinical Sciences, Marion duPont Scott Equine Medical Center, Virginia Tech, Leesburg, VA, USA*

ROBERT ZWEIGERDT • *Leibniz Research Laboratories for Biotechnology and Artificial Organs (LEBAO), Hannover Medical School, Hannover, Germany*

Methods in Molecular Biology (2016) 1502: 1–19
DOI 10.1007/7651_2016_335
© Springer Science+Business Media New York 2016
Published online: 14 April 2016

Multicompartmental Hollow-Fiber-Based Bioreactors for Dynamic Three-Dimensional Perfusion Culture

Eva Schmelzer and Jörg C. Gerlach

Abstract

The creation of larger-scale three-dimensional tissue constructs depends on proper medium mass and gas exchange, as well as removal of metabolites, which cannot be achieved in conventional static two-dimensional petri dish culture. In cultures of tissue-density this problem can be addressed by decentral perfusion through artificial micro-capillaries. While the static medium exchange in petri dishes leads to metabolite peaks, perfusion culture provides a dynamic medium supply, thereby preventing non-physiological peaks. To overcome the limitations of conventional static two-dimensional culture, a three-dimensional perfusion bioreactor technology has been developed, providing decentral and high-performance mass exchange as well as integral oxygenation. Similar to organ systems in vivo, the perfusion with medium provides nutrition and removes waste metabolites, and the perfusion with gas delivers oxygen and carbon dioxide for pH regulation. Such bioreactors are available at various dimensions ranging from 0.2 to 800 mL cell compartment volumes (manufactured by StemCell Systems, Berlin, Germany). Here, we describe in detail the setup and maintenance of a small-scale 4-chamber bioreactor with its tubing circuit and perfusion system.

Keywords: Bioreactor, Perfusion culture, Tissue engineering, Three-dimensional culture, Cell culture, Bioengineering, Hollow-fiber bioreactor

1 Introduction

Survival of cells within a tissue depends on proper supply with nutrients and oxygen as well as metabolic waste removal, which in vivo is enabled mostly by passive diffusion between blood vessels and is approximately limited to a distance of about 1 mm [1]. In contrast to conventional static two-dimensional petri dish culture, three-dimensional perfusion cultures provide sophisticated systems for the creation of larger-scale functional tissue constructs. The creation of such constructs depends on proper mass exchange of medium, oxygenation, and metabolite removal, which cannot be achieved in conventional static two-dimensional petri dish culture. Medium exchange in dishes is typically static, leading to non-physiological concentrations of metabolites, while in perfusion culture medium supply is dynamic, thereby avoiding metabolite peaks. Similar to organ systems in vivo, the perfusion with medium provides nutrition and removes waste metabolites, and perfusion with gas delivers oxygen and provides carbon dioxide for pH regulation. In the natural tissue, these tasks are accomplished by well-distributed arterial- and venous micro-capillaries. Correspondingly, in tissue-density culture, a perfusion of the tissue with artificial micro-capillaries can address these issues.

We have developed a bioreactor technology platform that is scalable because of an interwoven array of independent capillary systems (manufactured by StemCell Systems, Berlin, Germany), providing comparable mass exchange in smaller as well as larger bioreactors. The hollow-fiber based four-compartment bioreactors are available in various dimensions ranging from laboratory-scale (0.2 mL) to clinical-scale (800 mL), in which cells are cultured in a perfused three-dimensional environment. The clinical-scale bioreactor has been successfully used in a device to provide extracorporeal liver support in patients [2–4], providing approximately 600 g of liver neo-tissue. In addition, others and we have implemented different smaller-scale bioreactors as an advanced culture model with various primary cell types derived from different organs such as fetal or adult liver, bone marrow, or cord blood, as well as with cell lines and embryonic stem cells [5–20]. Each bioreactor is integrated into a processor-controlled device (Fig. 1) with

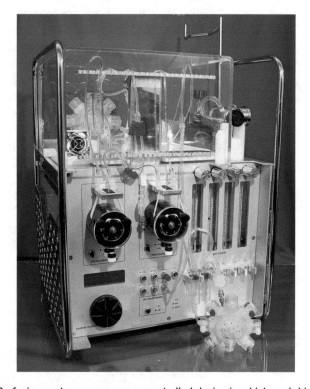

Fig. 1 Perfusion system, a processor-controlled device in which each bioreactor is integrated in.

electronic flow/pressure controlled perfusion, medium and waste pump operation, as well as heating and gas supply. An external tubing circuit for perfusion with medium and gas, respectively, is required.

A bioreactor contains two bundles of hydrophilic hollow-fiber microfiltration membranes for transport of culture medium, interwoven with one bundle of hydrophobic hollow-fiber oxygenation membranes for transport of oxygen and carbon dioxide (forming a gas compartment). The fibers are secured and potted within a polyurethane [21] housing, and cells are cultured in the interstitial spaces between the fibers (the cell compartment). The microfiltration fibers have a molecular weight cut off of MW 400,000 (Daltons), allowing larger proteins to pass through the fiber walls and into the cell compartment [21]. Culture medium circulates from the lumens of the microfiltration fibers to the cell compartment and back to the fiber lumens, due to the axial pressure drop from the inlet to the outlet of each fiber lumen (Starling flow) [21, 22]. The distribution of matter and perfusion circuit concepts have been described in more detail elsewhere [23, 24]. Medium is pumped through the two-microfiltration fiber bundles in opposing directions (counter-current flow), allowing the medium entering the cell compartment from one bundle (at its high pressure end) to exit by reentering the same bundle (at its low pressure end) or by entering the other bundle (at its low pressure end, adjacent to the first bundle's high pressure end). This complex flow pattern mimics an "arterial and venous" flow within natural tissues. Additionally, the interwoven oxygenation fibers ensure that the cells receive adequate oxygen, and carbon dioxide is regulated [21]. The three-dimensional nature of the cell compartment allows cells to form spontaneously tissue-like structures [25, 26] similar to those found in vivo. The convection-based mass transfer [26] and the mass exchange in the cell compartment allow formation of neo-endothelium, which enable physiologic perfusion and flow/pressure alterations as in parenchymal organs such as liver. For example, in culture of liver cells, tissue-like structures were spontaneously formed within 2–3 days [26]. These structures included neo-sinusoids with neo-formations of spaces of Dissé lined by endothelial cells, and structures resembling the Canals of Hering, the anatomical stem cell niche of liver progenitor cells. The vascular-like perfusion resulted in long-term support of a cell mass under substantial high-density conditions.

Here, we describe in detail the setup, filling procedure, and culture management of a smaller-scale, 4-chamber bioreactor, its tubing circuit, and the bioreactor perfusion system. The setup and maintenance of all laboratory-scale bioreactors is in principle similar, but slight differences exist in the kind of tubing systems to be used, gas flow rates, and speed of pumps for medium recirculation and feed.

2   Materials

Before beginning the process of assembling a complete 4-Chamber Analytic Scale Bioreactor with its associated perfusion circuit, the following supplies should be gathered. Similar substitutes of a different brand can be used in place of the specific products listed if necessary:

- 1 Sterile, surgical gown [Barrier Standard Repellent Surgical Gown with Tie Back and Sterile Towel—Fisher Scientific (P/N: 17-989-17F)].

- 2 Sterile, disposable sleeves [Cardinal Health Sterile Disposable Sleeves—Fisher Scientific (P/N: 19-120-3021)].

- 1 Pair of sterile, disposable, surgical gloves [Regent Medical Biogel Powder-Free Latex Surgical Gloves—Fisher Scientific— Size 6 (P/N: 18-999-4593B), Size 61/2 (P/N: 18-999-4593C), Size 7 (P/N: 18-999-4593D), Size 71/2 (P/N: 18-999-4593E), Size 8 (P/N: 18-999-4593F), Size 81/2 (P/N: 18-999-4593H), Size 9 (P/N: 18-999-4593G)].

- 2 Surgical masks [Kimberly Clark Technol Softtouch II Surgical Mask—Fisher Scientific (P/N: 18-999-4819)].

- 2 Sterile, surgical drapes [Andwin Scientific Sterile Drape Towels—18″ × 26″—Fisher Scientific (P/N: NC9560848)].

- 1 Pack of 9, sterile, dual-function, Luer-Lock caps [Smiths Medical (P/N: Blue—65834, Red—65833)].

- 1 Sterile, 4-Chamber Analytic Scale Bioreactor [StemCell Systems, Berlin, Germany].

- 1 Bioreactor perfusion circuit [StemCell Systems, Berlin, Germany].

- 4 6-in. extension tubing sets [Smiths Medical (P/N: MX452FL)].

- 2 20-in. extension tubing sets [Smiths Medical (P/N: MX450FL)].

- 1 33-in. extension tubing sets [Smiths Medical (P/N: MX451FL)].

- 3 48-in. extension tubing sets [Smiths Medical (P/N: MX048)].

- 4 Four-way stopcocks [Smiths Medical (P/N: MX9341L)].

- 4 Male-to-male Luer adapter [Smiths Medical (P/N: MX493)].

- 1 Gas perfusion condensation collection/flow indication tubing set [StemCell Systems, Berlin, Germany].

- 4 Injection caps [SAI Infusion Technologies P/N: IC].

- 1 0.2 μm hydrophobic gas filter [Qosina (P/N: 28213)].

- 1 0.2 μm hydrophilic medium sampling filter [Sartorius Minisart 0.2 μm Cellulose Acetate Membrane Syringe Filter with Luer-Lock—VWR International (P/N: 22002-046)].
- 1 Bioreactor perfusion system pump device [StemCell Systems, Berlin, Germany].

3   Methods

3.1   Setup of a 4-Chamber Analytic Scale Bioreactor

Ideally, the assembly of a complete bioreactor with its associated perfusion circuit is performed by two people: a "clean" person and a "sterile" person. In this manner, the clean person aseptically opens and passes components to the sterile person while maintaining the sterile integrity of each component. When connecting the tubing systems to the bioreactor, it is necessary to perform the work in a sterile air work-bench.

3.1.1   Connecting the Tubing System for Culture Medium Recirculation

1. Open the pack of 9, sterile, dual-function, Luer-lock caps on the sterile field and place one on each of the ports of the open ports of the bioreactor.
2. Open the 4-Chamber Analytic Scale Bioreactor sterile pack and place the bioreactor on the sterile field.
3. Place a dual-function, Luer-lock cap on each open port of the bioreactor.
4. Open a Universal Bioreactor Perfusion Circuit, place it on the sterile field, and connect tubes as described in the following text and according to Fig. 2
5. Remove the protective caps of ports 3 and 5.
6. Connect a 6″ extension tubing set to port 3 and a separate one to port 5.
7. Connect the two shorter red-striped tubes that end with male Luer-lock fittings to the 6″ extensions already in place on ports 3 and 5.
8. Remove the protective caps of ports 2 and 6.
9. Connect a 6″ extension tubing set to port 2 and a separate one to port 6.
10. If desired, connect an Intra-Circuit Culture Medium Temperature Measurement Assembly to the end of the 6″ extension of either port 2 or 6.
11. Connect the two blue-striped tubes with male Luer-locks to the 6″ extensions already in place on ports 2 and 6.
12. Place an injection cap to each of the ports of the four cell compartment chambers.

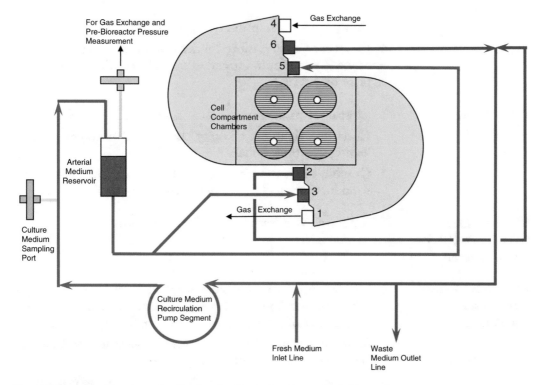

Fig. 2 Schematic of medium circuit setup for the 4-chamber analytic bioreactor

13. Insert a 20″ extension tubing set into the circuit at the Luer-lock connection situated after the arterial medium reservoir.

14. Insert a 33″ extension tubing set into the circuit at the Luer-lock connection situated before the waste medium outlet line.

15. If necessary, connect a 48″ extension tubing set to the end of the fresh culture medium supply pump tubing segment.

16. If necessary, connect a 20″ extension tubing set to the end of the 48″ extension tubing set.

17. If necessary, connect a 20″ extension tubing set to the end of the one-way valve of the waste medium outlet line.

18. Connect a sterile, hydrophilic filter to the small length of yellow tubing for culture medium sampling located between the recirculation pump tubing segment and the arterial medium reservoir.

3.1.2 Connecting the Tubing System for Mixed Gas Perfusion

1. For gas inlet, remove the protective caps of port 4.

2. Connect a sterile, hydrophobic gas filter to port 4.

3. Connect a 48″ extension tubing set to the sterile, hydrophobic, gas filter on port 4.

4. For gas outlet, remove the protective caps of port 1.

5. Connect a 48″ extension tubing set to port 1.

6. Connect a Gas Perfusion Condensation Collection/Flow Indication Tubing Set to the end of the 48″ extension set already in place on port 1.

3.1.3 Final Preparations and Inspections

1. Close all remaining open connections with Luer-lock caps.

2. Check the tightness of every connection thoroughly before leaving the sterile bench.

3. Carefully move the bioreactor and assembled perfusion circuit to an incubation chamber of a perfusion system.

3.2 Pre-culture Bioreactor Perfusion System Preparation

To be performed at least 1 day before cell inoculation.

3.2.1 General Bioreactor Perfusion System Preparations

1. Supply power to one bioreactor perfusion system. Supply power to additional bioreactor perfusion systems as needed per number of bioreactors to be used in the experiment.

2. Switch the main power, instrumentation lights, and heating elements to the "POWER ON" mode.

3. Calibrate all pumps that will be used during the experiment with their corresponding pump tubing segments (*see* **Note 1**)

4. After calibration and before mounting the bioreactors in the perfusion systems, all surfaces of the bioreactor perfusion systems and all pumps should be disinfected with a 70 % alcohol solution.

3.2.2 Mixed Gas Supply Assembly

1. Make sure that a compressed CO_2 tank has a sufficient amount of compressed gas within it to supply CO_2 for the entire experiment. For example, there must be at least 300 psi in a CO_2 tank (of about total capacity of 4500 psi) in order to run a two-bioreactor experiment for a period of 2 weeks.

2. Make sure the outlet stopcock of the gas flow indicator, the secondary reservoir of the Gas Perfusion Condensation Collection/Flow Indication Tubing Set, is in the "OPEN" position.

3. Connect the mixed gas supply to the bioreactor as described in the following text and according to Fig. 3.

4. Connect the braided, reinforced, silicone tubing coming from a building compressed air valve or dual-stage regulator to the port labeled "Air 2 In" using the attached stainless steel female Luer.

5. Connect the braided, reinforced, silicone tubing coming from a building compressed CO_2 valve or cylinder regulator to the port labeled "CO_2 In" using the attached stainless steel female Luer (*see* **Note 2**).

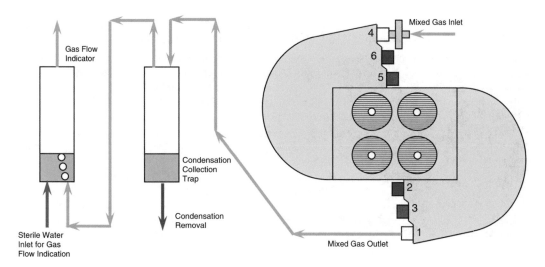

Fig. 3 Schematic of gas tubing setup for the 4-chamber analytic bioreactor

6. Connect a 6″ extension tubing set (Medex—Part # MX452FL) to the port labeled "Mix Out."

7. Connect a 3-Way Stopcock to the end of the 6″ extension tubing set already in place on the "Mix Out" port, such that the connection is made with the female Luer port of the stopcock that is perpendicular to the other two ports.

8. Connect an additional 6″ extension tubing set from the open female Luer port of the stopcock to the port labeled "Air 1 In."

9. Connect an additional 6″ extension tubing set from the open male Luer-lock port of the stopcock to one of the two open ports at the base of a 25 mL water resistance reservoir.

10. The port labeled "Compressor" will not be used during normal benchtop operation of the perfusion system and can be capped.

11. Connect a 30 mL syringe to the remaining open port at the base of the 25 mL water resistance reservoir.

3.2.3 Mixed Gas Supply Settings and Adjustment

Before filling the bioreactor, the gas lines have to be connected and the valves of the compressed air and CO_2 sources have to be turned to the "ON" position. Gas capillaries always need a positive pressure. The gas should always be connected and never turned off during a bioreactor perfusion experiment.

1. Make sure that the stopcock connecting the "Mix Out" port with the "Air 1 In" port and the 25 mL water resistance reservoir is open to all ports.

2. Using the 30 mL syringe attached to the 25 mL water resistance reservoir, fill the reservoir approximately ½ full with water.

Table 1
Suggested mixed gas supply operating flow rates for analytic scale bioreactors

Bioreactor model	Number of gas fibers	Standard operating total gas flow rate (mL/min)	Per fiber gas flow rate (mL/min)
8-mL lab scale	1456	40	0.027
4-chamber analytic scale	50	2.0	0.029
4-chamber analytic scale	50	2.5	0.036
4-chamber analytic scale	50	3.0	0.043
4-chamber analytic scale	50	3.5	0.050
4-chamber analytic scale	50	4.0	0.057

3. Set the rotameter labeled "Air 2" on the bioreactor perfusion system to a flow rate of approximately 30–40 mL/min.

4. Set the rotameter labeled "CO_2" on the bioreactor perfusion system to a flow rate of approximately 2–3 mL/min.

5. Connect the Bioreactor Mixed Gas Perfusion Inlet Line to the port labeled "Air 1 Out."

6. Adjust the volume of water in the 25 mL water resistance reservoir until the gas flow rate in the rotameter labeled "Air 1" is at the desired level (*see* **Note 3**).

7. Comparative per-fiber gas flow rates for analytic scale bioreactors are shown in Table 1 as a suggestion for operating conditions.

3.2.4 Final Preparations

1. Connect the ventilation line of the arterial reservoir to the pre-bioreactor pressure transducer port of the bioreactor perfusion system.

2. Fill the gas flow indicator with 5–10 mL of sterile water using a syringe. Then remove sterile water with the same syringe until water level is even with the top of the sterile water inlet port. The syringe will begin to draw air when this level is achieved. This step is necessary because if two bioreactors are being supplied with gas from the same perfusion system, the resistance to air flow must be the same through both bioreactor gas pathways. Therefore, their gas flow indicators must be filled with exactly the same volume of sterile water.

3. Make sure that the gas can easily pass through the bioreactor. If gas is passing through the bioreactor, gas bubbles should appear in the gas flow indicator of the Gas Perfusion Condensation Collection/Flow Indication Tubing Set. If not, check the gas outlet tubing set for kinks and make sure the gas flow

indicator outlet stopcock is in the "OPEN" position. If the gas outlet tubing set is kinked, blocked, or closed, the gas capillaries in the bioreactor may burst!

4. Connect the opposite end of the fresh culture medium feed line to the culture medium feed bottle and an empty, sterile, waste bottle to the opposite end of the waste medium outlet line.

5. Make sure that the stopcock on the ventilation port of all fresh culture medium and waste medium bottles is open to the bottle and the filtered port. This provides continuous ventilation of the bottles in order to compensate for the volume of withdrawn fresh culture medium/discharged waste medium.

3.2.5 Filling the System with Medium: General Preparations

1. Supply a sufficient amount of culture medium, at least 500 mL.

2. Open all clamps of the fluid circuit.

3. Place the bioreactor in a horizontal position.

4. Interrupt the fluid circuit between the fresh culture medium inlet line and the waste medium outlet line with a clamp.

5. Make sure that all Luer-lock connectors are securely fastened, as they tend to loosen on their own over time.

6. Check for gas bubbles in the sterile water of the gas flow indicator.

7. Insert the recirculation pump-tubing segment into the recirculation pump such that the outlet side goes to the arterial medium reservoir.

8. Do not insert the recirculation pump-tubing segment into the fresh culture medium feed pump.

9. Leave the halogen lamp heating system turned on during the filling procedure (the system needs to be at 37 °C ± 1 °C before the pressure can be checked).

3.2.6 Filling of the Tubing System

Before starting the recirculation pump, and thus the flow of culture medium into the system, it is important to properly adjust its settings. Turn the recirculation pump toggle switch to the "OFF" position.

1. By rotating the black dial, move the asterisk cursor until it appears next to the word "Rec Pump" on the blue digital display screen then press the dial like a button in order to be able to adjust the flow-rate setting.

2. By rotating the black dial, adjust the flow rate to 2 mL/min (*see* **Note 4**).

3. Start the recirculation pump by setting its toggle switch to the "ON" position.

4. Visually follow the pass of the culture medium from the bottle along the circuit, have your hand on the pump switch to guarantee quick stop if you see any problem (*see* **Note 5**).

5. Operate the recirculation pump in a start/stop fashion as needed.

6. When the culture medium reaches the arterial medium reservoir, stop the recirculation pump, place a clamp on the outlet line of the arterial medium reservoir, and open the ventilation line of the arterial medium reservoir by removing the clamp (if one is in place), and opening the stopcock to atmosphere.

7. Start the recirculation pump and fill the arterial medium reservoir approximately 3/5 full. When the arterial medium reservoir is filled to the desired level, stop the recirculation pump.

8. Re-clamp the ventilation line of the arterial reservoir, if necessary, re-close the stopcock to atmosphere, and remove the clamp on the outlet line of the arterial reservoir to maintain the desired liquid level.

3.2.7 Filling of the Cell Compartment and Capillary Membrane Bundles

1. Close the red clamp on the inlet line to the cell compartments. This will either be located at the end of a four-valve stopcock manifold or a two-valve stopcock manifold, depending on the circuit configuration used for the experiment.

2. Start the recirculation pump to fill the bioreactor culture medium inlet lines, and subsequently, the bioreactor cell compartment and capillary membrane bundles. Slow filling at a pump speed of 2 mL/min or less should result in the absence of gas bubbles in the cell compartment from the start.

3. Visually observe the cell compartments as they fill with culture medium to check for gas bubbles (*see* **Note 6**).

4. After this is completed, continue running the recirculation pump and allow culture medium to fill the waste medium outlet line.

5. Continue to run the recirculation pump until all gas bubbles have been removed from the bioreactor and the venous side of the recirculation circuit. When/if this is completed, stop the recirculation pump.

If the standard filling procedure cannot fill the entire recirculation circuit and cell compartment with culture medium, it may be necessary to use the following procedure to assist in the filling of the bioreactor:

1. Open the red clamp on the port 3 inlet line while closing the red clamps of all other inlet lines.

2. Close the blue clamps on the port 2 outlet line and the cell compartment outlet line while leaving the blue clamp on the port 6 outlet line open.

3. Start the recirculation pump and hold the flow-head of port 3 in a downward position to release trapped gas bubbles from the capillary membrane bundle.

4. When visual inspection reveals no more gas bubbles passing through the capillary membrane bundle, stop the recirculation pump.

5. Open the red clamp on the port 5 inlet line while closing the red clamp on the port 3 inlet line.

6. Open the blue clamp on the port 2 outlet line while closing the blue clamp on the port 6 outlet line.

7. Start the recirculation pump and hold the flow-head of port 5 in a downward position to release trapped gas bubbles from the capillary membrane bundle.

8. When visual inspection reveals no more gas bubbles passing through the capillary membrane bundle, stop the recirculation pump.

9. Open the red clamp on the port 3 inlet line and open the blue clamp on the cell compartment outlet line while closing the blue clamp on the port 2 outlet line.

10. Start the recirculation pump and rotate the bioreactor in order to mobilize any trapped gas in the cell compartments. When it appears that gas bubbles have ceased to exit the cell compartment, stop the recirculation pump (*see* **Note 7**).

11. With the blue clamp on the cell compartment outlet line still open, open the red clamp on the cell compartment inlet line while closing the red clamps on the port 3 and 5 inlet lines.

12. Start the recirculation pump and allow culture medium to pass through the stopcock valve manifold in order to mobilize any gas bubbles from the cell compartment to the venous side of the circuit. When it appears that gas bubbles have ceased to exit the stopcock valve manifold, stop the recirculation pump.

13. Fasten the fresh culture medium feed pump tubing segment in the fresh culture medium feed pump and lock the raceway in place.

14. Remove and/or release the clamp between the fresh culture medium feed line and the waste medium outlet line, allowing culture medium to recirculate.

3.2.8 Initial Recirculation

1. Ensure that all clamps of the fluid circuit are open.

2. Start the fresh culture medium feed pump and set it to a flow rate of 0.5 mL/h. Verify that the fresh culture medium feed pump is producing forward flow by visually observing the waste bottle.

Table 2
Suggested operating culture medium recirculation flow rates for analytic scale bioreactors

Bioreactor model	Number of perfusion fibers	Standard operating total gas flow rate (mL/min)	Per fiber gas flow rate (mL/min)
8-mL lab scale	840	30	0.036
4-chamber analytic scale	50	3.0	0.060
4-chamber analytic scale	50	4.0	0.080
4-chamber analytic scale	50	5.0	0.100
4-chamber analytic scale	50	6.0	0.120
4-chamber analytic scale	50	7.0	0.140
4-chamber analytic scale	50	8.0	0.160
4-chamber analytic scale	50	9.0	0.180
4-chamber analytic scale	50	10.0	0.200

3. Start the recirculation pump and increase the recirculation flow rate gradually to 7 mL/min.

4. Comparative per-fiber culture medium flow rates for 4-chamber analytic scale bioreactors are shown in Table 2 as a suggestion for operating conditions.

5. After 60 min of recirculation, check for gas bubbles in the bioreactor and recirculation circuit. If necessary, use gentle manipulations and light tapping of the bioreactor and recirculation circuit to remove the gas bubbles (*see* **Note 8**).

6. If culture medium is detected in the gas lines or capillaries at any time during the initial recirculation phase, there is a leak in the gas capillaries and the bioreactor cannot be used for cultivation.

7. If possible, culture medium should be circulated for approximately 24 h before cell injection.

3.3 Bioreactor Perfusion System Cell Injection

At the day of the start of the experiment, before cell injection, it is important to make sure that the bioreactor environment is optimal for cell maintenance and/or proliferation.

3.3.1 Pre-cell Injection Culture Management

1. Tighten all Luer-lock connections as needed.

2. Check the incubation chamber temperature and adjust as needed. It is recommended that the bioreactor be kept in the bioreactor perfusion system incubation chamber with an ambient temperature of approximately 37 °C ± 1 °C (*see* **Note 9**).

3. Stop the recirculation pump.

4. Place a clamp between the culture medium sampling port and the arterial medium reservoir.

5. Open the ventilation line of the arterial reservoir by removing and or releasing the clamp, if necessary, and opening the stopcock to atmosphere.

6. Place a sterile, disposable 1 mL syringe in the Luer port of the culture medium sample port filter.

7. Turn on the recirculation pump until the syringe fills with approximately 1 mL of culture medium.

8. Turn off the recirculation pump.

9. Place a second sterile, disposable 1 mL syringe in the Luer port of the culture medium sample port filter.

10. Turn on the recirculation pump until the syringe fills with approximately 0.7 mL (or as much as needed) of culture medium.

11. Turn off the recirculation pump.

12. Use the culture medium to analyze baseline parameters of the cell culture medium as desired, such as pH, pCO_2, pO_2, glucose, and lactate.

13. Adjust the pH by adjusting the CO_2 flow rate, as necessary. Increasing the CO_2 flow rate will lower the pH and vice versa. The physiological pH range for most cell types is 7.35–7.45.

3.3.2 Pre-cell Injection Air Removal Process

If culture medium is recirculated for a long period of time, gas bubbles may have collected in the cell compartment. Additionally, gas bubbles resulting from the increase in culture medium temperature may appear and accumulate in the recirculation circuit.

1. Tighten all Luer-lock connections as needed.

2. Operate the recirculation pump in a start/stop fashion as needed.

3. Verify that the arterial reservoir is approximately 3/5 full and that all clamps of the fluid circuit are open.

4. Visually inspect the bioreactor cell compartment and recirculation circuit for gas bubbles.

5. Attempt to work the gas bubbles to the arterial reservoir using gentle manipulations and light tapping of the bioreactor and recirculation circuit (*see* **Note 10**).

6. It may be necessary to pass medium through each perfusion fiber pathway, and the cell compartment, individually, as described before, to assist in the removal of air bubbles.

3.3.3 Cell Injection Procedure

Before cell injection, ensure that the injected cell volume does not exceed the bioreactor cell compartment volume. Each cell compartment of a four-chamber bioreactor has an approximate volume of 0.18 mL, for a total cell compartment volume of 0.72 mL. Prepare cells in medium within a syringe with needle of appropriate diameter. Prepare a second syringe with needle filled with medium only.

1. Disinfect the membrane of the injection cap with 70 % alcohol.

2. Inject the cells gently through the membrane of the injection cap.

3. To flush cells completely into the cell chamber, subsequently inject medium through the membrane of the injection cap.

4. During injection do not to allow air to enter the bioreactor cell compartment.

5. Close the ventilation line of the arterial reservoir by clamping, if necessary, and closing the stopcock.

3.4 Bioreactor Perfusion System Culture Management

3.4.1 Culture Management

Best results for operating the 4-chamber analytic scale bioreactor were achieved with the following conditions:

- Clamps on port 2, 3, 5, and 6 lines are open.
- Clamps on cell compartment inlet and outlet lines are closed.
- Stopcock of cell injection line(s) is/are closed to all ports.
- Fresh culture medium inlet and waste medium outlet lines are unclamped.
- Ventilation line of the arterial reservoir is clamped and/or closed.
- Pre-bioreactor pressure measurement is enabled.
- Perfusion mode: Cross flow.
- Recirculation rate: 3–10 mL/min.
- Fresh medium feed-rate: 0.5–2.0 mL/h, according to the cells' needs.
- Compressed air flow rate: ~3.63 mL/min per bioreactor.
- Compressed CO_2 flow rate: ~0.36 mL/min per bioreactor.
- Culture temperature: 37 °C ± 1 °C in the bioreactor chamber.
- Medium pH: 7.35–7.45, according to the cells' needs.

3.4.2 Air Removal/Cell Compartment Flush Procedure

Do not initiate this air removal/cell compartment flush procedure involving cell compartment outflow, when non-adherent cells are used. Additionally, do not open the cell compartment outlet after inoculating cells through culture day 3, as the cells may flow out of the cell compartment. Perform the gas bubble removal procedure only as necessary by visual inspection of the cell compartments, after the cells have attached. This is a test of bioreactor integrity. It is also potentially necessary since gas bubbles may be very slowly transferred through the oxygenation membranes and may accumulate in the cell compartment, injuring the cells.

1. Make sure that all Luer-lock connectors are securely fastened, as they tend to loosen on their own over time.

2. Operate the recirculation pump in a start/stop fashion as needed.

3. Adjust the level of the arterial reservoir to 3/5 full with medium as needed.

4. If not already clamped, clamp the cell compartment inlet line while opening all clamps of the fluid circuit, including the clamp on the cell compartment outlet line.

5. Visually inspect the bioreactor cell compartment and recirculation circuit for gas bubbles.

6. Attempt to work the gas bubbles to the arterial reservoir by gentle manipulations of the bioreactor and circuit. When it appears that there are no gas bubbles in the bioreactor and gas bubbles have ceased to exit the cell compartment, close the clamp on the cell compartment outlet line (*see* **Note 10**).

4   Notes

1. The integrated pumps of the perfusion system for recirculation and feed should be calibrated before starting a new experiment. It is critical to use the optimal contact pressure when inserting the tube into the pump to avoid inaccuracies between the actual and the displayed value. The most common cause for malfunction is setting the contact pressure too high. To check the contact pressure, feed the pump tube into the raceway and place the output of the tube into a container with water. Then replace the raceway and turn the setting screw underneath the lever to the left until there is no more forward flow, no bubbling in the water. Now carefully turn the setting screw to the right until you see air bubbles forming in the water. To secure the feed turn the screw an additional ¼ turn to the right. Do not overwind the setting screw. Excessive contact pressure can lead to reduction or blockage of the flow, compression of the tube in front of the clam jaw at the output, premature deterioration or breakage of the tube, or can cause excessive resistance and slow down/overstress the pump motor.

2. If desired during benchtop operation and during perfusion system transport, the braided, reinforced, silicone-tubing segment for compressed CO_2 coming from the bioreactor perfusion system may be attached to the outlet port of the building

compressed CO_2 valve or the outlet port of a compressed CO_2 cylinder regulator. This compressed CO_2 inlet line is equipped with a one-way valve that disables leakage of CO_2 through it when compressed CO_2 is being perfused through the "CO_2 In" Luer port. However, when using this line to perfuse compressed CO_2, it is necessary to cap the "CO_2 In" Luer port to prevent leakage of CO_2 through it.

3. Keep in mind that increasing the volume of water in the reservoir will increase the gas flow rate to the bioreactor.

4. When using a 4-chamber analytic scale bioreactor, it is not necessary to use the 1.0 mm ID recirculation pump tubing segment in place of the standard 2.5 mm ID recirculation pump tubing segment in order to achieve lower flow rates.

5. It may be necessary to temporarily increase the recirculation pump speed to start the flow of culture medium out of the feed bottle.

6. The 4-chamber analytic scale bioreactor is fragile and cannot be vigorously shaken to remove gas bubbles. Gentle manipulations and light tapping of the bioreactor and recirculation circuit may aid in the removal of gas bubbles.

7. This procedure also tests the flow through the capillary membrane walls of both perfusion pathways.

8. It may be necessary to pass medium through each perfusion fiber pathway, and the cell compartment, individually, to assist in the removal of air bubbles.

9. The incubation chamber temperature is only displayed on the temperature control unit of the first-generation perfusion systems (with three pumps). The second-generation perfusion systems (with two pumps) do not have a user adjustable temperature control unit. The control unit is built in to the internal circuitry and maintains the incubation chamber at $37\,^{\circ}\mathrm{C} \pm 1\,^{\circ}\mathrm{C}$ via control of the halogen lamps.

10. Removal of gas bubbles from the bioreactor and recirculation circuit may result in a decrease in fluid volume of the arterial reservoir. Refill as necessary.

Acknowledgements

This work was supported by NIH grant 5R01HL108631 and by the University of Pittsburgh Medical Center. The authors thank Daniel McKeel for his continuous excellent technical support.

References

1. Pittman RN (2011) The circulatory system and oxygen transport. In: Regulation of tissue oxygenation. Morgan & Claypool Life Sciences, San Rafael, CA. doi: 10.4199/C00029ED1V01Y201103ISP017

2. Busse B, Smith MD, Gerlach JC (1999) Treatment of acute liver failure: hybrid liver support. A critical overview. Langenbecks Arch Surg 384(6):588–599

3. Sauer I, Zeilinger K, Pless G, Kardassis D, Theruvath T, Pascher A, Mueller A, Steinmueller T, Neuhaus P, Gerlach J (2003) Extracorporeal liver support based on primary human liver cells and albumin dialysis—treatment of a patient with primary graft nonfunction. J Hepatol 39(4):649–653

4. Sauer IM, Kardassis D, Zeillinger K, Pascher A, Gruenwald A, Pless G, Irgang M, Kraemer M, Puhl G, Frank J, Muller AR, Steinmuller T, Denner J, Neuhaus P, Gerlach JC (2003) Clinical extracorporeal hybrid liver support-phase I study with primary porcine liver cells. Xenotransplantation 10:460–469

5. Gerlach JC, Brayfield C, Puhl G, Borneman R, Muller C, Schmelzer E, Zeilinger K (2010) Lidocaine/monoethylglycinexylidide test, galactose elimination test, and sorbitol elimination test for metabolic assessment of liver cell bioreactors. Artif Organs 34(6):462–472, doi: AOR885 [pii] 10.1111/j.1525-1594.2009.00885.x

6. Gerlach JC, Hout M, Edsbagge J, Bjorquist P, Lubberstedt M, Miki T, Stachelscheid H, Schmelzer E, Schatten G, Zeilinger K (2009) Dynamic 3D culture promotes spontaneous embryonic stem cell differentiation in vitro. Tissue Eng Part C Methods 16(1):115–121, doi: 10.1089/ten.TEC.2008.0654

7. Gerlach JC, Lubberstedt M, Edsbagge J, Ring A, Hout M, Baun M, Rossberg I, Knospel F, Peters G, Eckert K, Wulf-Goldenberg A, Bjorquist P, Stachelscheid H, Urbaniak T, Schatten G, Miki T, Schmelzer E, Zeilinger K (2010) Interwoven four-compartment capillary membrane technology for three-dimensional perfusion with decentralized mass exchange to scale up embryonic stem cell culture. Cells Tissues Organs 192(1):39–49, doi: 000291014 [pii] 10.1159/000291014

8. Gerlach JC, Over P, Foka HG, Turner ME, Thompson RL, Gridelli B, Schmelzer E (2014) Role of transcription factor CCAAT/enhancer-binding protein alpha in human fetal liver cell types in vitro. Hepatol Res 45 (8):919–932. doi:10.1111/hepr.12420

9. Pekor C, Gerlach JC, Nettleship I, Schmelzer E (2015) Induction of hepatic and endothelial differentiation by perfusion in a three-dimensional cell culture model of human fetal liver. Tissue Eng Part C Methods. doi:10.1089/ten.TEC.2014.0453

10. Schmelzer E, Mutig K, Schrade P, Bachmann S, Gerlach JC, Zeilinger K (2009) Effect of human patient plasma ex vivo treatment on gene expression and progenitor cell activation of primary human liver cells in multi-compartment 3D perfusion bioreactors for extra-corporeal liver support. Biotechnol Bioeng 103(4):817–827. doi:10.1002/bit.22283

11. Schmelzer E, Triolo F, Turner ME, Thompson RL, Zeilinger K, Reid LM, Gridelli B, Gerlach JC (2010) Three-dimensional perfusion bioreactor culture supports differentiation of human fetal liver cells. Tissue Eng Part A 16 (6):2007–2016. doi:10.1089/ten.TEA.2009.0569

12. Stachelscheid H, Wulf-Goldenberg A, Eckert K, Jensen J, Edsbagge J, Bjorquist P, Rivero M, Strehl R, Jozefczuk J, Prigione A, Adjaye J, Urbaniak T, Bussmann P, Zeilinger K, Gerlach JC (2013) Teratoma formation of human embryonic stem cells in three-dimensional perfusion culture bioreactors. J Tissue Eng Regen Med 7(9):729–741. doi:10.1002/term.1467

13. Hoffmann SA, Muller-Vieira U, Biemel K, Knobeloch D, Heydel S, Lubberstedt M, Nussler AK, Andersson TB, Gerlach JC, Zeilinger K (2012) Analysis of drug metabolism activities in a miniaturized liver cell bioreactor for use in pharmacological studies. Biotechnol Bioeng 109(12):3172–3181. doi:10.1002/bit.24573

14. Mueller D, Tascher G, Muller-Vieira U, Knobeloch D, Nuessler AK, Zeilinger K, Heinzle E, Noor F (2011) In-depth physiological characterization of primary human hepatocytes in a 3D hollow-fiber bioreactor. J Tissue Eng Regen Med 5(8):e207–e218. doi:10.1002/term.418

15. Housler GJ, Miki T, Schmelzer E, Pekor C, Zhang X, Kang L, Voskinarian-Berse V, Abbot S, Zeilinger K, Gerlach JC (2012) Compartmental hollow fiber capillary membrane-based bioreactor technology for in vitro studies on red blood cell lineage direction of hematopoietic stem cells. Tissue Eng Part C Methods 18 (2):133–142. doi:10.1089/ten.TEC.2011.0305

16. Housler GJ, Pekor C, Miki T, Schmelzer E, Zeilinger K, Gerlach JC (2014) 3-D perfusion bioreactor process optimization for CD34+ hematopoietic stem cell culture and differentiation towards red blood cell lineage. J Bone

Marrow Res 2:3. doi:10.4172/2329-8820. 1000150

17. Darnell M, Schreiter T, Zeilinger K, Urbaniak T, Soderdahl T, Rossberg I, Dillner B, Berg AL, Gerlach JC, Andersson TB (2011) Cytochrome P450-dependent metabolism in HepaRG cells cultured in a dynamic three-dimensional bioreactor. Drug Metab Dispos 39(7):1131–1138. doi:10.1124/dmd.110. 037721

18. Miki T, Ring A, Gerlach J (2011) Hepatic differentiation of human embryonic stem cells is promoted by three-dimensional dynamic perfusion culture conditions. Tissue Eng Part C Methods 17(5):557–568. doi:10.1089/ten. TEC.2010.0437

19. Ring A, Gerlach J, Peters G, Pazin BJ, Minervini CF, Turner ME, Thompson RL, Triolo F, Gridelli B, Miki T (2010) Hepatic maturation of human fetal hepatocytes in four-compartment three-dimensional perfusion culture. Tissue Eng Part C Methods 16 (5):835–845. doi:10.1089/ten.TEC.2009. 0342

20. Zeilinger K, Schreiter T, Darnell M, Soderdahl T, Lubberstedt M, Dillner B, Knobeloch D, Nussler AK, Gerlach JC, Andersson TB (2011) Scaling down of a clinical three-dimensional perfusion multicompartment hollow fiber liver bioreactor developed for extracorporeal liver support to an analytical scale device useful for hepatic pharmacological in vitro studies. Tissue Eng Part C Methods

17(5):549–556. doi:10.1089/ten.TEC.2010. 0580

21. Gerlach JC, Encke J, Hole O, Muller C, Ryan CJ, Neuhaus P (1994) Bioreactor for a larger scale hepatocyte in vitro perfusion. Transplantation 58(9):984–988

22. Gerlach J, Schnoy N, Encke J, Müller C, Smith M, Neuhaus P (1995) Improved hepatocyte in vitro maintenance in a culture model with woven multicompartment capillary systems: electron microscopy studies. Hepatology 22:546–552

23. Gerlach JC, Witaschek T, Strobel C, Brayfield CA, Bornemann R, Catapano G, Zeilinger K (2010) Feasibility of using sodium chloride as a tracer for the characterization of the distribution of matter in complex multi-compartment 3D bioreactors for stem cell culture. Int J Artif Organs 33(6):399–404

24. Balmert SC, McKeel D, Triolo F, Gridelli B, Zeilinger K, Bornemann R, Gerlach JC (2011) Perfusion circuit concepts for hollow-fiber bioreactors used as in vitro cell production systems or ex vivo bioartificial organs. Int J Artif Organs 34(5):410–421. doi:10.5301/ijao. 2011.8366

25. Sussman NL, Kelly JH (1993) Improved liver function following treatment with an extracorporeal liver assist device. Artif Organs 17 (1):27–30

26. Gerlach J, Neuhaus P (1994) Culture model for primary hepatocytes. In Vitro Cell Dev Biol 30A:640–642

Methods in Molecular Biology (2016) 1502: 21–33
DOI 10.1007/7651_2016_336
© Springer Science+Business Media New York 2016
Published online: 01 April 2016

A Bioreactor to Apply Multimodal Physical Stimuli to Cultured Cells

Jan-Christoph Edelmann*, Lizzie Jones*, Remi Peyronnet, Liang Lu, Peter Kohl, and Ursula Ravens

Abstract

Cells residing in the cardiac niche are constantly experiencing physical stimuli, including electrical pulses and cyclic mechanical stretch. These physical signals are known to influence a variety of cell functions, including the secretion of growth factors and extracellular matrix proteins by cardiac fibroblasts, calcium handling and contractility in cardiomyocytes, or stretch-activated ion channels in muscle and non-muscle cells of the cardiovascular system. Recent progress in cardiac tissue engineering suggests that controlled physical stimulation can lead to functional improvements in multicellular cardiac tissue constructs. To study these effects, aspects of the physical environment of the myocardium have to be mimicked in vitro. Applying continuous live imaging, this protocol demonstrates how a specifically designed bioreactor system allows controlled exposure of cultured cells to cyclic stretch, rhythmic electrical stimulation, and controlled fluid perfusion, at user-defined temperatures.

Keywords: Bioreactor, Excitable cells, Cyclic stretch, Electrical stimulation, Perfusion, Cell mechanics, Electrophysiology, Myocardium

---

1 Introduction

The physical environment that cells experience in the cardiac niche influences many cellular functions. For instance, cyclic stretch has been shown to mediate growth factor secretion [1], and to alter calcium-handling in cardiomyocytes [2] and production of extracellular matrix proteins by cardiac fibroblasts [3, 4]. Similarly, isolated adult rat ventricular myocytes in primary culture subjected to continual electric field stimulation exhibit improved contractile properties and calcium-transient characteristics compared to unstimulated quiescent cells [5, 6]. These studies highlight the important roles of physical stimulation in sustenance of isolated cells and cultures.

Several "biomimetic" strategies have been applied to steer functional improvement of ex vivo cultivated 3D tissue constructs by mimicking aspects of the native myocardial environment.

---

*These authors contributed equally.

Stimuli applied include cyclic stretch [7], electrical pacing [8], "electro-tensile" stimulation (where electrical and mechanical stimuli are applied in controlled sequence) [9], constant fluid perfusion [10–12], oxygenation via carriers through channeled scaffolds [12], and more recently, concerted medium perfusion in combination with electrical stimulation [13].

Various bioreactors have been designed to implement some of the above stimulation strategies in cultured cells or tissue constructs. Generally, though, bioreactors are either cyclic stretch or electrical stimulation chambers. Our custom reactor combines mechanical and electrical stimulation in one apparatus. It allows one to study the interplay of well-coordinated uniaxial cyclic stretch and electrical stimulation. Additionally, it is possible to adapt the perfusate's composition, temperature, and flow rate. During experiments, the reactor is mounted on top of an inverted microscope for continuous imaging.

In the following, we describe the design and use of an improved version of the custom made bioreactor, first introduced in 2013 [14, 15] at the TU Dresden. We provide a step-by-step description of the protocols required to use this integrated system to study excitable and non-excitable cells.

2 Materials

2.1 List of Materials and Equipment

All parts with direct or indirect contact to the cells under investigation have to be sterilized before use (underlined in the list below). The assembly and loading of the reactor must be performed in a culture hood under sterile conditions. Subsequent experiments may be performed outside the hood in a reasonably clean room as usually maintained for tissue culture.

2.1.1 Experimental Setup

1. Equipment, devices and apparatus for standard tissue culture work.

2. Polydimethylsiloxane polymer substrate (PDMS) mold [Fortech UG, TU Dresden].

3. Bioreactor, composed of the assembled chamber and driving unit [Fortech UG, TU Dresden]. Add-ons to the setup to seal the chamber [homebrew].

4. External electrical stimulation control box, cables, and carbon electrodes. Photoelectric Sensor and cables [all Fortech UG, TU Dresden].

5. Motor control box and cables [Fortech UG, TU Dresden] to apply cyclic, uniaxial stretch.

6. 250 mL glass bottle with neck thread [Simax] and watertight lid with 4 sealable cap-ports [Schott]. All parts together serve as solution reservoir.

7. $3\times$ 2 mL serological pipette tips [Sarsted] for medium handling and gas bubbling.

8. Exit port with a serological pipette tip [Sarsted] and siphon [homebrew] for outward gas handling.

9. Carbogen (5 % CO_2, 95 % O_2) [BOC Medical] with 22 μm millipore filter [Merck Millipore] attached.

10. Carbogen sparger [BEOT].

11. Peristaltic pump [Gilson] for medium perfusion (rate up to 5 mL/s).

12. $2\times$ Silicone tubes [Saint-Gobain], 6 mm outer diameter, 3 mm inner diameter, 75 and 100 cm long. The shorter tube is arranged between the medium reservoir and the peristaltic pump. The longer tube conducts carbogen gas into the reservoir.

13. $2\times$ Silicone tubes [Saint-Gobain], 4 mm outer diameter, 3 mm inner diameter, 75 cm long each. The first tube is connected to the peristaltic pump and serves as forward supply for the chamber. The second is used as reverse supply.

14. 40 °C water bath with pump [Haake] for water jacketing and to heat the medium in the solution reservoir.

15. A fixation stage [homebrew] to stabilize the reservoir.

16. 3 silicone tubes [Saint-Gobain], 6 mm outer diameter, 3 mm inner diameter, 50 cm long each. Tubes are thermally insulated and used to supply water to the water jacket.

17. Flow gauge indicator [Bürkle].

18. Tubing connectors [World Precision Instruments].

2.1.2 Data Acquisition

1. Inverted microscope [Olympus] equipped with long distance objectives $5\times$, $10\times$ and $20\times$ [Thalheim Spezial Optik] and an adapted microscope table [Fortech UG, TU Dresden].

2. Microscope Camera [Nikon].

3. Computer Workstation [HP] with Data Acquisition Software [Fortech UG, TU Dresden].

2.2 General Description of the Setup

The microscope stage can accommodate up to four bioreactors that can be used in parallel [14, 15]. Each of the reactors consists of two parts: The bio-chamber and the driving unit (Fig. 1). The chamber is perfused with culture medium and houses two stretchable PDMS membranes onto which cells of interest are seeded (*see* **Note 1**). Two carbon electrodes are fitted in the chamber and allow for pacing with defined temporal off-set to the stretch protocol. Cell responses to a set of user-defined, ambient stimuli are monitored via long-working distance objectives and a camera connected to a

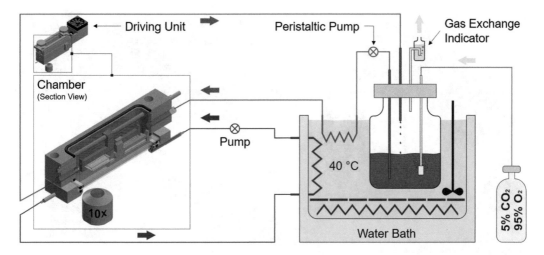

Fig. 1 General schematic of the setup. The bioreactor, composed of the driving unit and the chamber (left hand side), is connected to a water bath for temperature control and to a bottle of culture medium for cell perfusion. This bicarbonate buffered solution is bubbled with carbogen to keep the pH constant. The bioreactor's fan, stepper motor, pacing electrodes, and photoelectric sensor are connected to their external control boxes (not shown)

computer for bright field time-lapse imaging. All materials used for the construction of the chamber meet demands on biocompatibility and sterilization. Stainless steel is used for rods, screws, the bottom framework and the lid. Silicone rubber is used for sealing inlets and outlets and connections between microscope glass slides and the bioreactor housing. Polyether ether ketone (PEEK) polymer was selected for all other parts.

A 250 mL glass bottle containing 100 mL of suitable, bicarbonate buffered culture medium is held in a water bath set to 40 °C (Fig. 1). A peristaltic pump, located between the water bath and the chamber of the bioreactor (*see* **Note 2**), circulates the culture medium at a flow rate of 0.5–1.0 mL/s, to maintain the nutrient and gas supply in a vital range for the cells in the bioreactor. The culture medium is bubbled in its reservoir with carbogen to keep pH constant (*see* **Note 3**). An exit tube with a siphon resting in ethanol (*see* **Note 4**) ensures that there is no contamination from ambient atmosphere and indicates the gas exchange. Thermally insulated tubing conveys water from a temperature-controlled water bath to the fluid perfused lower section of the chamber (henceforth referred to as "water jacket") in a closed loop configuration. Its purpose is to supply the top chamber with thermal energy to compensate for heat losses to the environment. A flow gauge (enclosed propeller not shown in Fig. 1) next to the bioreactor indicates water jacket's flow (*see* **Note 5**).

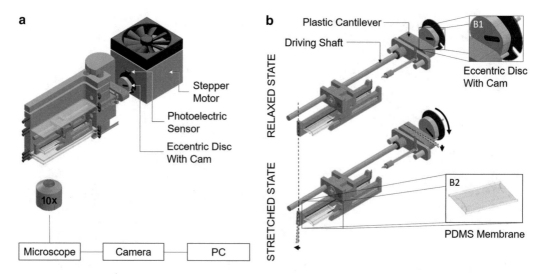

Fig. 2 Key components of the bioreactor. (**a**) Section view of the chamber connected to the driving unit located above a microscope's objective. The subsequent signal acquisition chain is sketched underneath. (**b**) View of the inside components of the chamber allowing cyclic, uniform uniaxial stretch. The eccentric disk with cam (**B1**) rotates and acts on the PEEK cantilever which transmits the movement to a shaft that drives the left membrane holder away from the right, fixed one. This introduces stretch to the membrane. (**B2**) Focus on a carefully designed PDMS membrane which is designed to compensate for Poisson's effect through arc-like reinforcements on both sides

2.3 Stretching

Torque from a stepper motor (Fig. 2a) is converted into a uniaxial stretch pattern via an eccentric disk with cam (Fig. 2a, B1) that acts on a PEEK cantilever (Fig. 2b). This cantilever drives a shaft with one set of the PEEK brackets that hold the stretchable membranes. The second set of brackets, arranged in parallel, is fixed. The stretchable PDMS membrane is clamped between the two neighboring brackets (Figs. 2 and 3a#3). Thus, rotational motion transmitted by the cantilever translates into a lateral change in distance between the parallel edges of the two brackets (Fig. 2b). The membrane, clamped between the brackets, experiences pseudo-uniaxial deformation, which is transferred to the cells attached to it. To improve homogeneity of strain, the short sides of the membrane have arc-shaped indentations with reinforced edges (Fig. 2B2), which compensate for (some of) the Poisson's effects (*see* **Note 6**). The stretch, stretch/relaxation velocities and frequency are highly adaptable by adjusting the settings of the stepper motor (0.25–4 Hz), while stretch waveform and amplitude can be altered by exchanging the eccentric disk (0–20 % ratio of deformation, Fig. 2B1). A standard protocol used involves 10 % stretch at 1 Hz for 24 h. Modifying the shape of the eccentric disk adjusts the mechanical deformation behavior of the setup: Smooth or steep slopes before and after the tip of the cam allow one to accurately control stretch and relaxation kinetics. Transitions can be adjusted to match physiological cardiac deformations in this ex vivo system. On the other hand, increasing the diameter of the cam and creating a "dip" instead of a "bump" allows one to maintain cell stretch and apply a short relaxation, mimicking a different mechanical loading pattern. Stretch pattern options are infinite and easily adjustable.

2.4 Pacing

Homogeneous electrical stimulation is applied via carbon electrodes (Fig. 3a#4) positioned in close proximity to the membrane holding brackets. The pattern of electrical stimulation is flexible by changing, in software, parameters for rate (for example to match mechanical protocols), pulse duration (1–10 ms), current amplitude (1–10 mA) and timing. Temporal interrelation of electrical and mechanical stimulation is controlled via photoelectric feedback from a sensor located on the axis of the motor (Fig. 2a).

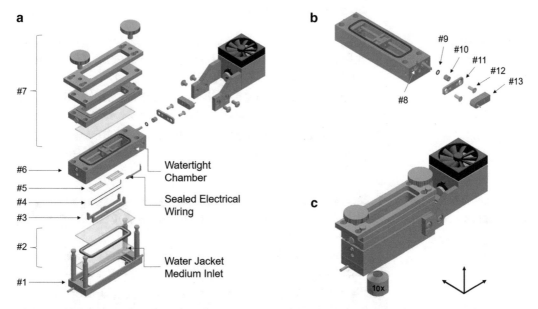

Fig. 3 Bioreactor assembly. (**a**) Exploded view of the chamber for step by step assembly. (#1) Stainless steel constituting the frame of the chamber and allowing water circulation for the water jacket. (#2) The water jacket. The machined microscope glass slide with two holes serving as inlet and outlet for the water jacket at the bottom (the left hole is hidden). The inside of the water jacket is framed to the sides by the spacer surrounded by a rubber seal. It is closed by an off-the-shelf microscope glass slide. (#3) Membrane holder clamps. (#4) Carbon electrodes and wiring. (#5) PDMS Membranes. (#6) Main part of the chamber hosting the shaft which transmits the motor's energy to the membrane holders. (#7) Top glass that seals the chamber and screw top allowing controlled tightening of the whole setup. (**b**) Depiction of the seal of the driving shaft. (#8) Wiring outlet for the carbon electrodes. (#9) 5 mm O-Ring. (#10) PEEK shell. (#11) PEEK plate. (#12) M3, steel screws. (#13) PEEK cantilever. (**c**) The complete bioreactor ready to be used in isometric representation. *Arrows* serve as scale reference where each *arrow* indicates a length of 3 cm

3 Methods

3.1 PDMS Membrane Preparation

3.1.1 Molding

1. Make a solution of silicon: Dow Corning Sylgard 184; base–curing agent ratio = 10:1.

2. Pour this solution into a custom-made polycarbonate casting mold in a clean room environment (*see* **Note 7**).

3. Leave the mold undisturbed for 72 h at room temperature to allow curing.

4. Store resulting PDMS substrates (2 cm × 1 cm × 50 mm) in a petri dish to prevent from dust contamination (*see* **Note 8**).

The following steps are to be performed under a standard cell culture hood to work in very clean, preferably sterile, conditions.

3.1.2 Sterilization

1. Incubate PDMS membrane in 70 % ethanol for 30 min (*see* **Note 9**).

2. Wash 2 times with sterile Phospahte-Buffered Saline (PBS).

3. Leave to dry for 10 min.

3.1.3 Coating

1. Coat PDMS substrates with a 1:100 mixture of gelatin, type I collagen, fibronectin, or collagen-fibronectin. Use water as dilution medium (*see* **Note 10**).

2. After 4 h, wash PDMS surfaces three times with PBS and seed cells of interest on coated substrates (*see* **Note 11**).

3. After 12 h, visually inspect cell attachment to PDMS substrates. Wash samples three times with warm PBS (37 °C) to remove non-adherent cells.

3.2 Bioreactor Construction

1. Sterilize all parts of the bioreactor using a steam autoclave (121 °C, 200 kPa, 20 min) or incubating in 70 % ethanol for 30 min (*see* **Note 12**).

2. If using 70 % ethanol for sterilization, all parts need to be dried entirely in the hood.

3. Place the two 4 mm circular rubber rings (O-rings) and one 60 mm O-ring in the framework of the bioreactor to ensure a water tight seal can be created later for the water jacket (Fig. 3a#1).

4. Next, insert the glass slide with the two circular openings for the water jacket, followed by the spacer with the rubber seal and the microscope glass slide (Fig. 3a#2).

5. Take the central chamber's component (Fig. 3a#6) hosting the two brackets. At this point, their clamps are not attached yet. Next, fit the carbon electrodes (Fig. 3a#4) to the brackets with medical suture material, ensuring antiparallel

current flow (*see* **Note 13**). Make sure the wiring is fitted firmly into its opening from the inside of the chamber. The glued construction is depicted in Fig. 3a#4 while Fig. 3b#8 illustrates its assembled position (*see* **Note 14**). Afterwards, the clamps (Fig. 3a#3) are clipped onto their corresponding brackets. Each combination of bracket and clamp will serve as a membrane holder later. However, PDMS membranes (Fig. 3a#5) are not attached yet. Subsequently, prepare to seal the slide bearing by adding the 5 mm O-ring (Fig. 3b#9), the PEEK shell (Fig. 3b#10), and PEEK plate (Fig. 3b#11). Fasten the screws (Fig. 3b#12) in order to squeeze the O-ring into the surrounding construction until the opening is watertight. A good trade-off has to be found that ensure the friction levels for the driving shaft remain low. Fine tune the screws' position in **step 8**) (*see* **Note 15**).

6. Check that both the bottom and top of the central chamber are equipped with a 60 mm O-ring each. Afterwards, the prepared central chamber's component is added to the preassembled bottom framework parts.

7. Add the second glass slide to complete the chamber, followed by a PEEK spacer and two metal lids in ascending size to allow the unit to be screwed closed (Fig. 3a#7).

8. Circulate sterile PBS to flush any remaining ethanol. Adjust and assess the sealing of the driving shaft (*see* **Note 15**). Once sorted, add and fasten the PEEK cantilever (Fig. 3b#13).

9. Circulate with pre-warmed (37 °C) culture medium to check for lack of leakages (*see* **Note 16**).

10. Stop the circulation, open the top of the chamber, and remove the culture medium from the chamber.

11. Take out the central part of the chamber, flip it upside down, and mount the membranes (pre-seeded with cells of interest) on the holders (*see* **Note 17**).

12. Repeat **steps 6** and **7** and while circulating culture medium mount the driving unit. The bioreactor is now complete (Fig. 3c).

13. Fix the bioreactor to the stage of the inverted microscope.

14. Connect the chamber's water jacketed to the water bath (*see* **Note 18**).

15. Connect the motor, the electrodes, and the photoelectric sensor to their corresponding control boxes for driving and coordinated pacing.

At this point, experiments can proceed for periods of up to 4 days (or more).

3.3 Improvements from the Original Design

One major amendment of the set-up addresses the way in which the measurement chamber is encapsulated. Previously, gas exchange between the culture medium in the chamber and the surrounding incubator environment was promoted. Ventilation was possible through the feed openings for the driving shaft and the pacing electrode connection wires. For long term experiments, these openings are prone to contamination, once the chamber is taken from the sterile incubator environment onto the microscope stage, as well as culture medium leakage. Thus, wires are glued into a shell (Fig. 3a) and gas exchange with the environment is suppressed, once the chamber is fully assembled (Fig. 3b#8). The slide bearing is equipped with a 5 mm O-ring from the outside of the chamber to seal the opening (Fig. 3b#9). As the drive shaft needs to move easily, a shell (Fig. 3b#10), plate (Fig. 3b#11) and screws (Fig. 3b#12) are used to secure the ring's position from the outside. Via the screws, holding pressure is adjusted such that the chamber is sealed but the shaft still able to move with low friction levels.

Another improvement addresses changes in the culture medium solution supply line. Previously, the solution was kept in a fridge, warmed up and used once before put to waste. We changed solution handling to a closed loop strategy in order to reduce costs and to be able to run long-term experiments, for example to test effects of (at times expensive) drugs.

To maintain the pH of the DMEM based culture medium (bicarbonate buffered) we bubble the perfusion solution in its reservoir with carbogen. To visually control gas exchange rate and avoid untoward pressure gradients in the system, gas is allowed to exit through a siphon filled with ethanol (Fig. 1).

Furthermore, a holder for the reservoir bottle to be kept in place in the water bath has been added. This ensures upright alignment of the tubing at a reproducible height throughout and between experiments (not shown).

In addition to coordinated stretch and electrical stimulation, it is possible to grow cells on glass coverslips which can be placed in the central chamber's component (Fig. 3a#6) section of the bioreactor, above the water jacket. This allows live imaging while pacing and controlling perfusion, solution composition, temperature, and/or flow for short or long-term experiments.

Combining this setup with a motorized computer-controlled microscope stage allows acquisition of time lapse image series from different regions of interest and different incubators. This, together with an increase in the number of bioreactors used in parallel, would potentially allow use as a high-throughput system.

In the future, we intend to improve the shape of the chamber, to apply controlled fluid shear to cells cultured on PDMS membranes. Applying both shear and stretch while pacing and perfusing would open new avenues for research relevant for cardiovascular applications and provide a better understanding of the role of a wider range of physical stimuli on cells.

4   Notes

1. PDMS is used due to its ease of molding, low cytotoxicity, oxygen permeability, and transparency. Each bioreactor chamber can house 2 PDMS substrates at a time (Figs. 1, 2a and 3a#5) thus all experiments can be performed in duplicate or one membrane can host control cells and the other mutant/treated cells for comparison within the same experiment.

2. After leaving the pump, the tubing is dipped in the water bath to compensate for heat loss occurring at the level of the pump. The tubing section in between the water bath and bioreactor is kept as short as possible to avoid heat loss. The tubing used in this section has a small open diameter in order to increase the flow rate in this section and thus reduce the time for heat exchange. Temperature maintenance is further improved by application of insulating material around that tube. The pump is installed upstream of the chamber to generate a slightly positive pressure in the chamber when feeding culture medium. Having the pump downstream of the chamber would create a slightly negative pressure which would increase the contamination risk if the chamber is not fully airtight. Having a slightly positive pressure increases the risk of perfusate leaks, but this should not prevent the experiment from continuing (assuming the leak is not massive); this is in contrast to the alternative scenario (solution contaminated), which would end experiments in all simultaneously perfused chambers. Finally, hydrostatic pressure should be considered when the system is installed. We found that the tendency for the chamber to leak was increased when the level of the bioreactor was about 20 cm lower relative to the perfusate level in the reservoir; this should be avoided.

3. A Millipore filter is applied before the carbogen gas bubbles the solution to ensure sterile conditions.

4. To avoid excessive pressure in the system, the tubes exit point should be close to the liquid surface (volume of ethanol between 1 and 5 mL). A positive pressure in the reservoir avoids infiltration and contamination from outside pathogens.

5. The propeller indicator needs adjustment and fixation such, that the inlet and outlet are aligned horizontally at the top of the device. That way, small air bubbles are transported further into the reservoir directly. This prevents gas accumulation in the indicator and, as a consequence, ensures functionality.

6. Poisson's effect is leading to PDMS membrane shortening perpendicular to the stretch direction. This effect is compensated for by the membranes due their sophisticated design. Validation was done with ARAMIS surface studies as presented in Refs. [14, 15].

7. Pay close attention to not get any dust on the PDMS membranes. Dust particles are extremely hard to remove once attached.

8. To reduce the risk of contamination, it is recommended to seal the dish externally with Parafilm.

9. Activation of the membranes' surface via plasma treatment can be done but is not considered a mandatory procedure. In general, activation is used to treat materials that are normally hydrophobic. After treatment with plasma, they become more hydrophilic and more susceptible to chemical modifications or coating with proteins. However, adhesion of cells after coating both the ozone plasma treated membranes and raw membranes yielded comparable results in our hands when growing cardiac fibroblasts.

10. We found that PDMS membranes had to be coated as our cells of interest (fibroblasts and cardiomyocytes) did not adhere otherwise.

11. As a positive control, seed the same numbers of cells in a standard 24-well plate. Both the culture plate and the PDMS substrates have an effective growth area of 2 cm^2.

12. Sterilization is a crucial step. The bioreactor is made of numerous parts and if the entire assembly is not perfectly clean, contamination will occur. Contamination is not only prone to waste of resources and time, but may yield misleading results.

13. The carbon electrodes have to be tied to the PEEK brackets, and electrically connected in an antiparallel fashion, to allow for a more homogeneous electrical field distribution [14]. The electrodes need to be clamped to the brackets facing the inside of the chamber.

14. Fitting the wiring appropriately is crucial for sealing and thus obtaining a watertight chamber. Fluid tightness is important to avoid leakages and contamination.

15. In order to avoid leakage and spill of PBS in the lab, the adjustment of the screws should be done by using negative pressure in a hood in sterile conditions. Fill up the chamber of the fully assembled reactor with PBS via a syringe. Once all air is removed, block the outlet of the main chamber. Connect a 2 mL syringe to the inlet and apply negative pressure to the chamber by pulling on it. The appearance of air bubbles inside the chamber indicates that the chamber is leaking. The position of the leak can be identified by visual inspection. If this happens at the opening for the driving shaft, the screws have to be fastened further. Perform multiple iterations. It is advised to apply and hold negative pressure via the syringe for a few seconds and rotate the driving shaft. Fine-tune the

adjustment such that a good trade-off between friction level and water tightness is found. This adjustment is crucial for the success of the experiment.

16. Leakages are usually the result of incorrectly mounted parts or missing rubber O-rings. The mounting step is crucial as it is a precondition for the entire experiment. Cell culture medium is used to assess the setup, as the solution is colored and leakages are spotted more easily than using PBS.

17. When attaching the PDMS membranes, the eccentric disk shall be fastened to avoid accidental movements and damage to the cells. Instead of PDMS membranes, other deformable constructs, tissue sections, or trabecular muscle samples may be studied.

18. To ensure cells are kept within a suitable temperature range during the entire experiment, the water bath and pump shall be switched on well before cells are put into the chamber (minimum 20 min earlier). The bypass can be used before the chamber is installed on the microscope's table to circulate warm water through the tubing. This allows the system to warm up and reduce the time needed to reach the target temperature range later, once the fluid perfused section of the chamber is connected.

Acknowledgement

This work has been supported by the European Research Council Advanced Grant *CardioNECT*, and the Magdi Yacoub Institute at Harefield. RP is holder of an Imperial College Junior Research Fellowship; PK is a Senior Fellow of the British Heart Foundation. We acknowledge Heinz-Felix Körber and Eva Rog-Zielinska for technical assistance.

References

1. Leychenko A, Konorev E, Jijiwa M, Matter ML (2011) Stretch-induced hypertrophy activates NFkB-mediated VEGF secretion in adult cardiomyocytes. PLoS One 6(12): e29055. doi:10.1371/journal.pone.0029055

2. Tsai CT, Chiang FT, Tseng CD, Yu CC, Wang YC, Lai LP, Hwang JJ, Lin JL (2011) Mechanical stretch of atrial myocyte monolayer decreases sarcoplasmic reticulum calcium adenosine triphosphatase expression and increases susceptibility to repolarization alternans. J Am Coll Cardiol 58(20):2106–2115. doi:10.1016/j.jacc.2011.07.039

3. Husse B, Briest W, Homagk L, Isenberg G, Gekle M (2007) Cyclical mechanical stretch modulates expression of collagen I and collagen III by PKC and tyrosine kinase in cardiac fibroblasts. Am J Physiol Regul Integr Comp Physiol 293(5):R1898–R1907. doi:10.1152/ajpregu.00804.2006

4. Lee AA, Delhaas T, McCulloch AD, Villarreal FJ (1999) Differential responses of adult cardiac fibroblasts to in vitro biaxial strain patterns. J Mol Cell Cardiol 31(10):1833–1843. doi:10.1006/jmcc.1999.1017

5. Berger HJ, Prasad SK, Davidoff AJ, Pimental D, Ellingsen O, Marsh JD, Smith TW, Kelly RA (1994) Continual electric field stimulation preserves contractile function of adult ventricular myocytes in primary culture. Am J Physiol 266 (1 Pt 2):H341–H349

6. Holt E, Lunde PK, Sejersted OM, Christensen G (1997) Electrical stimulation of adult rat cardiomyocytes in culture improves contractile properties and is associated with altered calcium handling. Basic Res Cardiol 92 (5):289–298

7. Zimmermann WH, Melnychenko I, Eschenhagen T (2004) Engineered heart tissue for regeneration of diseased hearts. Biomaterials 25(9):1639–1647

8. Radisic M, Park H, Shing H, Consi T, Schoen FJ, Langer R, Freed LE, Vunjak-Novakovic G (2004) Functional assembly of engineered myocardium by electrical stimulation of cardiac myocytes cultured on scaffolds. Proc Natl Acad Sci U S A 101(52):18129–18134. doi:10.1073/pnas.0407817101

9. Feng Z, Matsumoto T, Nomura Y, Nakamura T (2005) An electro-tensile bioreactor for 3-D culturing of cardiomyocytes. A bioreactor system that simulates the myocardium's electrical and mechanical response in vivo. IEEE Eng Med Biol Mag 24(4):73–79

10. Carrier RL, Rupnick M, Langer R, Schoen FJ, Freed LE, Vunjak-Novakovic G (2002) Perfusion improves tissue architecture of engineered cardiac muscle. Tissue Eng 8(2):175–188. doi:10.1089/107632702753724950

11. Radisic M, Yang L, Boublik J, Cohen RJ, Langer R, Freed LE, Vunjak-Novakovic G (2004) Medium perfusion enables engineering of compact and contractile cardiac tissue. Am J Physiol Heart Circ Physiol 286(2):H507–H516. doi:10.1152/ajpheart.00171.2003

12. Radisic M, Marsano A, Maidhof R, Wang Y, Vunjak-Novakovic G (2008) Cardiac tissue engineering using perfusion bioreactor systems. Nat Protoc 3(4):719–738. doi:10.1038/nprot.2008.40

13. Maidhof R, Tandon N, Lee EJ, Luo J, Duan Y, Yeager K, Konofagou E, Vunjak-Novakovic G (2012) Biomimetic perfusion and electrical stimulation applied in concert improved the assembly of engineered cardiac tissue. J Tissue Eng Regen Med 6(10):e12–e23. doi:10.1002/term.525

14. Lu L, Mende M, Yang X, Korber HF, Schnittler HJ, Weinert S, Heubach J, Werner C, Ravens U (2013) Design and validation of a bioreactor for simulating the cardiac niche: a system incorporating cyclic stretch, electrical stimulation, and constant perfusion. Tissue Eng Part A 19(3–4):403–414. doi:10.1089/ten.TEA.2012.0135

15. Lu L, Ravens U (2013) The use of a novel cardiac bioreactor system in investigating fibroblast physiology and its perspectives. Organogenesis 9(2):82–86. doi:10.4161/org.25014

Methods in Molecular Biology (2016) 1502: 35–52
DOI 10.1007/7651_2015_312
© Springer Science+Business Media New York 2015
Published online: 13 December 2015

Aggregate and Microcarrier Cultures of Human Pluripotent Stem Cells in Stirred-Suspension Systems

Preeti Ashok, Yongjia Fan, Mahboubeh R. Rostami, and Emmanuel S. Tzanakakis

Abstract

Pluripotent stem cells can differentiate to any cell type and contribute to damaged tissue repair and organ function reconstitution. The scalable culture of pluripotent stem cells is essential to furthering the use of stem cell products in a wide gamut of applications such as screening of candidate drugs and cell replacement therapies. Human stem cell cultivation in stirred-suspension vessels enables the expansion of stem cells and the generation of differentiated progeny in quantities suitable for use in animal models and clinical studies. We describe methods of culturing human pluripotent stem cells in spinner flasks either as aggregates or on microcarriers. Techniques for assessing the quality of the culture and characterizing the cells based on the presentation of pertinent markers are also presented. Spinner flask culture with its relatively low capital and operating costs is appealing to laboratories interested in scaling up their production of stem/progenitor cells.

Keywords: Pluripotent stem cells, Differentiation, Scalable culture, Xeno-free culture, Spinner flasks, Cell aggregates, Microcarriers

1 Introduction

Pluripotent stem cells with their capacity to differentiate into a broad range of cell types [1] are a promising source of cellular material for clinical applications and screening platforms in drug development. Human pluripotent stem cells (hPSCs) are successfully used in cell replacement studies and animal models after their differentiation to retinal pigment epithelium [2] in humans, spinal cord cells [3], neural cells to treat ischemia stroke in rats [4], and endothelial-like cells to treat hind limb ischemia in mice [5]. Realization of stem cell-based therapies will require the development of scalable systems for the expansion of stem/progenitor cells and their differentiation to desired phenotypes. For example, a minimum of 1–2×10^9 cells is required for replacing damaged cells in the myocardium after infarction [6]. Given also the <100 % differentiation efficiencies, it becomes obvious that typical dish cultures are not ideal for generating differentiated cells for studies in animal models and eventually for clinical therapies.

Stirred-suspension bioreactors are an attractive modality for stem cell cultivation. While a fully automated benchtop bioreactor system affords tight control and real-time monitoring of several culture variables, the associated capital, maintenance, and operation costs as well as the required knowledge basis for its use make it a less suitable option for many laboratories. In this chapter, we detail methods for the culture of hPSCs (embryonic—hESCs and induced pluripotent stem cells—hiPSCs) in spinner flasks, which require minimal investment, are simple to operate, and provide a reasonable approximation of larger stirred-suspension systems.

Human stem cells can be cultured in spinner flasks as aggregates [7, 8], on microcarriers [9, 10] of after encapsulation [11]. In this chapter, we describe methods developed in the lab for the culture of stem cells in spinner flasks either as aggregates or on microcarriers. Aggregate cultures are simpler to set up and require less downstream processing than microcarrier cultures. The beads however provide a larger surface area-to-volume ratio, which may lead to more efficient use of culture media and factors. Moreover, the presence of concentration gradients is typically more pronounced in aggregate cultures, which may impose challenges, particularly for directed differentiation of stem cells in a spinner flask following their expansion. The characterization of cultured cells includes the detection of expressed pluripotency and/or differentiation markers by flow cytometry, immunohistochemistry, and quantitative polymerase chain reaction (qPCR). Karyotyping is also carried out to assess gross abnormalities caused during the course of successive passages. The differentiation capacity of the cells is tested by coaxing cell specification toward the three germ layers.

2 Materials

2.1 Static Culture of Pluripotent Stem Cells Using Vitronectin or Matrigel

1. Human embryonic or induced pluripotent stem cells [e.g., H9 hESCs (WiCell Research Institute) or any other hPSC line of interest].

2. Biosafety hood.

3. Tissue culture microscope with epifluorescence attachment and camera.

4. Tissue culture plates, e.g., 6-cm dishes, 6-well and 12-well plates (Corning).

5. Pipettes, pipet aids, and 1 ml, 200 μl, and 10 μl pipet tips.

6. E8 medium: Mix basal medium with 20× and 500× supplements along with penicillin/streptomycin (optional: if using, add 5 ml per 500 ml total medium) (Stemcell Technologies; available as 100× solution) (*see* **Note 2**).

7. Water bath (37 °C).

8. A tissue culture CO_2 incubator (37 °C; 5 % CO_2). A stirring table is placed in the incubator for spinner flask culture (Fig. 1a).

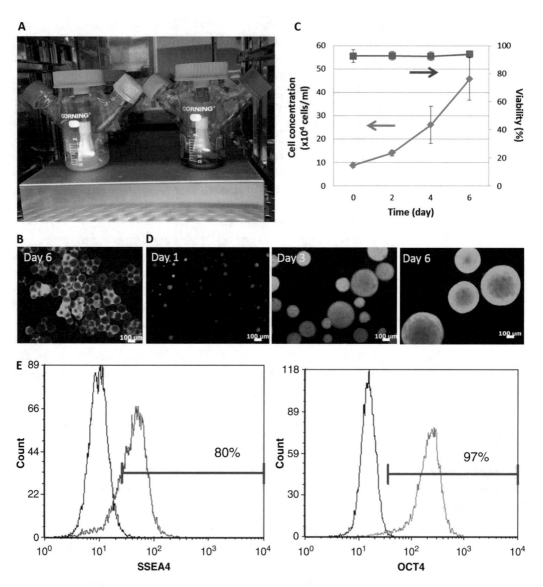

Fig. 1 Human pluripotent stem cell culture in spinner flasks. (**a**) Spinner flasks in a CO_2 incubator. Human H9 embryonic stem cells are cultured as aggregates in E8 medium (*left*). Spinner flasks shown on the right are prepared for passaging these cells. (**b**) Human H9 cells grown on Matrigel-coated polystyrene beads in a spinner flask. Bar: 100 µm. (**c**) Typical growth profile and viability of hPSCs grown as aggregates in a spinner flask. Each point is shown as mean ± standard deviation from triplicates. (**d**) Cell aggregates stained with FDA on days 1, 3, and 6 of culture. Bar: 100 µm. (**e**) Determination of the expression of pluripotency markers SSEA4 and OCT4 using flow cytometry for hPSCs after 6 days of stirred suspension culture. *Black curves*: Isotype control. *Colored curves*: Antigen. The fractions of positive cells may vary between experiments

9. Matrigel (Corning) (*see* **Note 1**)

10. Recombinant human vitronectin, truncated (Life Technologies)

11. Dispase (Life Technologies)

12. DMEM/F-12 (Life Technologies)

2.2 Microcarrier Culture of Stem Cells in Spinner Flasks

1. Phosphate buffer saline, free of Ca^{+2} and Mg^{+2} (PBS; Life Technologies).

2. Accutase (Innovative Cell Technologies).

3. SoloHill Fact III collagen-coated polystyrene microcarriers (Pall Corp.) for use after being coated with Matrigel.

4. SoloHill plastic microcarriers (Pall Corp.) for use after being coated with vitronectin.

5. ProCulture spinner flasks (125 or 250 ml; Corning).

6. Multiposition magnetic stirring plate (Thermo Scientific) with controller.

7. Matrigel solution.

8. Recombinant human vitronectin, truncated.

9. Human serum albumin (HSA; Sigma Aldrich).

10. Y-27632 Rho-associated protein kinase (ROCK) inhibitor (Enzo Biochem; 10 mM solution).

11. Pluronic F-68 (Life Technologies).

12. Fluorescein diacetate (FDA; Sigma Aldrich): Stock solution is prepared by dissolving the FDA powder in dimethylsulfoxide (DMSO) to a concentration of 10 mM; the stock solution is diluted to 10 μM (working concentration) with PBS.

13. 100-μm mesh strainer (BD Falcon).

14. Trypan Blue (Life Technologies).

15. Lactate dehydrogenase (LDH) cytotoxicity detection assay (Roche).

2.3 Differentiation to Definitive Endoderm and Mesoderm Cells in Spinner Flask Cultures

1. RPMI 1640 medium (Life Technologies).

2. CTS B-27 supplement, xeno-free (Life Technologies).

3. Recombinant human Activin A (R&D Systems).

4. Recombinant human BMP4 (R&D Systems).

5. Knockout serum replacer (KSR; Life Technologies).

2.4 Characterization

2.4.1 Total RNA Isolation, Reverse Transcription, and Quantitative PCR

1. Trizol (Life Technologies).

2. Isopropanol (100 % pure; HPLC grade).

3. Ethanol (200 proof; HPCL grade).

4. Chloroform (HPLC grade).

5. Nuclease-free water.

6. OligodT (Thermo Fisher).

7. ImProm-II 5× Buffer (Promega).

8. $MgCl_2$ (Promega).

9. dNTP mix (Thermo Fisher).

10. ImProm-II Reverse Transcriptase (Promega).

11. Dynamo SYBR Green mix set (Thermo Scientific).

2.4.2 Flow Cytometry

1. 4 % Paraformaldehyde solution: To make a 100 ml solution, mix 4 g paraformaldehyde (J.T. Baker) in 100 ml PBS and heat the mixture to 65 °C while stirring using a magnetic stirrer with heating option. Increase pH by adding NaOH until the solution clears (allow 2–4 h for solution to clear). The final pH is to be around 6.9 (*see* **Note 4**).

2. PBS.

3. Normal donkey serum (NDS; Jackson ImmunoResearch Laboratories, Inc.) or other blocking agent depending on the antibodies used.

4. TrypLE (Life Technologies).

5. Cytonin (Trevigen).

2.4.3 Immunohistochemistry

1. Paraformaldehyde (4 %) solution.

2. Bovine serum albumin (BSA; Sigma Aldrich).

3. Labtek 8-well microscope glass slides with surface area $0.7 \, cm^2$/well (Thermo Scientific).

4. Deionized (DI) water.

5. Octylphenylpolyethylene glycol or Triton X (J.T. Baker).

2.4.4 Karyotyping

1. T-75 flasks (Corning).

2. KaryoMAX Colcemid Solution (Gibco from Life Technology).

3. Cell hypotonic solution (CHS) is made by making a solution of 40 mM KCl, 20 mM HEPES, 0.5 mM EDTA, and 9 mM NaOH in DI water.

4. Acetic acid.

5. Methanol.

6. GelMount aqueous mounting medium (Sigma-Aldrich).

7. Wright-Giemsa stain, modified (Sigma-Aldrich).

2.4.5 Differentiation into Mesoderm

1. DMEM (Life Technologies).

2. Sodium selenite (Sigma Aldrich).

3. RPMI 1640 vitamin solution (100×) (Sigma-Aldrich).

4. Chemically defined (CD) lipid concentrate (Life Technologies; sold as 1000× solution).

5. RPMI 1640 amino acid solution (50×) (Sigma-Aldrich).

6. Holo-transferrin (Sigma-Aldrich).

7. MEM nonessential amino acids (100×) (Life Technologies).

8. Sodium pyruvate (Life Technologies; sold as 100× solution).

9. Penicillin/streptomycin.

10. Carrier-free recombinant human BMP4 (R&D Systems).

11. Carrier-free recombinant human WNT3A (R&D Systems).

12. KY02111 (Xcessbio Biosciences, Inc).

13. Matrigel or recombinant human vitronectin as substrate.

2.4.6 Differentiation to Ectoderm Cells

1. DMEM/F12 medium.

2. Neurobasal medium (Life Technologies).

3. B-27 supplement minus vitamin A (50×) (Life Technologies).

4. Glutamax (100×) (Life Technologies).

5. N2 supplement (100×) (Life Technologies).

6. FGF2 (R&D Systems).

7. Collagenase IV (Life Technologies).

8. Low binding non-treated polystyrene 6-cm dishes (BD Biosciences).

2.5 Antibodies

1. Anti human-NANOG (BD Pharmingen).

2. Anti-human OCT4 (Santa Cruz Biotechnology, Inc.).

3. Anti-human SSEA4 (Abcam).

4. Antihuman TRA-1-60 (BD Pharmingen).

5. Anti-human KDR (R&D Systems).

6. Anti-human PDGFRα (R&D Systems).

7. Anti-human C-KIT (R&D Systems).

8. Anti-human Nestin (R&D Systems).

9. Anti-human βIII-Tubulin (Sigma-Aldrich).

10. Appropriate secondary antibodies (e.g., from Jackson Immu-noResearch Laboratories, Inc.).

3 Methods

3.1 Static (Dish) Culture

Aseptic conditions must be maintained during cell culture and manipulations.

1. Two days before thawing cells, thaw Matrigel at 4 °C overnight. Prepare 400-μl aliquots in the biosafety hood taking care to use icepacks to keep Matrigel cold. Freeze the aliquots at −80 °C.

2. If using vitronectin, thaw the vial at room temperature, prepare 10-μl aliquots, and store them at −80 °C.

3. Prepare E8 medium aliquots in 50 ml tubes and store them at −20 °C.

4. On the day before thawing cells, thaw aliquots of Matrigel and E8 medium at 4 °C overnight.

5. Using ice-cold DMEM/F12, make a solution of Matrigel in DMEM/F12 (1:70). Coat each dish to be used immediately with the Matrigel solution. If using vitronectin, a solution of vitronectin in PBS at the ratio 1:100 should be prepared for coating dishes.

6. Warm E8 medium at room temperature. Remove the cryovial with the frozen cells from the liquid nitrogen tank and place it in the 37 °C water bath. Remove the vial from the bath when the last crystal of ice is remaining, spray with 70 % ethanol, and swipe with a paper towel.

7. Transfer the contents of the vial to a 15-ml tube in the biosafety hood using a 5-ml pipet.

8. Using a 10-ml pipet, add 9 ml of E8 medium to the tube dropwise while shaking so that the cells do not go under osmotic shock.

9. Centrifuge the tube at $200 \times g$ for 5 min. Meanwhile, aspirate the Matrigel solution from the dish and replace it with a volume of E8 medium that is 1 ml less than the total volume to be used.

10. Aspirate the medium from the centrifuge tube, and use a 1-ml pipette tip to flush the pellet at the bottom with E8 medium exactly once. Remove the contents using the pipette tip and add it slowly to the dish. Gently shake the dish to ensure even distribution of the colonies.

11. To passage cells on Matrigel, incubate them in dispase for 1–2 min. Wash them twice with DMEM/F12 and scrape them using a 10-ml glass pipet. Add the scraped colonies to a coated dish splitting the cells at ratios ranging between 1:4 and 1:6. If cells are on vitronectin, incubate them in 0.5 M EDTA solution for 1–2 min at room temperature. Wash the cells twice with PBS. Using a 1-ml pipette tip, flush the dish with medium. Seed the cells at ratios between 1:4 and 1:6.

3.2 Stirred-Suspension Culture in Spinner Flasks

3.2.1 Microcarrier Culture Coated with Vitronectin, HSA, and UV Radiation

1. Wash and autoclave the spinner flasks before use.

2. Weigh 0.5 g of plastic microcarriers suspended in PBS and autoclave them. The amount of beads can vary depending on the total surface of beads needed.

3. Remove as much of the PBS as possible without disturbing the beads.

4. Mix 200 μl vitronectin in 2 ml PBS, and add it to the beads. Mix the beads with the vitronectin solution. Incubate the beads in vitronectin for at least 1 h; resuspend the beads every 20 min.

5. Dissolve HSA in PBS at a concentration of 0.01 g/ml. Filter sterilize the solution. Add 2 ml of the albumin solution to the beads with vitronectin so that the final concentration of albumin is 0.005 g/ml. Incubate the beads in the solution for exactly 30 min.

6. Meanwhile, treat the cells intended for seeding with 10 μM ROCK inhibitor for 1 h.

7. Transfer the beads and the solution to a 35-mm sterile culture dish. Remove as much as 3 ml supernatant leaving behind just enough solution to allow the beads to move around. Tape the dish to a shaker, and gently shake the dish without the lid under UV light (254 nm) for 40 min.

8. Use E8 medium with 10 μM ROCK inhibitor to transfer the beads to a 15-ml tube. Wash the dish with the medium until you collect all the beads.

9. Let the beads settle down and aspirate as much of the medium as possible. Wash once more with the wash medium. Let the beads incubate in the wash medium until ready for use.

10. Add appropriate volume of Accutase (e.g., 2 ml Accutase for a 6-cm dish) to the cells ready to be seeded and incubate them for 10 min at 37 °C to disperse them into single cells. Harvest the cells by adding an equal volume of medium containing ROCK inhibitor. Spin down the cells at $200 \times g$ for 5 min. Count the cells using a cell counter.

11. Aspirate the wash medium without disturbing the beads and add fresh E8 medium with ROCK inhibitor so that the total volume including the cells and medium with beads equals 8 ml. Transfer the beads to a 10-cm petri dish.

12. Add ten million cells to the top of the beads.

13. Manually shake the dish a few times to distribute the cells among the beads and allow them to stick. Place them in the incubator, and manually shake the dish every 15 min for the first hour, and then every 30 min for the next 3 h. Add 4 ml of medium with 10 μM ROCK inhibitor and let the beads remain with the cells overnight in the incubator (37 °C, 5 % CO_2).

14. The next morning, transfer the beads to the spinner flask and add another 38 ml (total of 50 ml) medium containing 0.02 % Pluronic F-68 (diluted 1:500 from a 10 % Pluronic F-68 stock) (*see* **Note 3**).

15. Take a 1 ml sample and let the microcarrier with cells settle. Remove the supernatant, add 1 ml of 10 μM FDA solution, and incubate for 1 min. Check for cell attachment under a fluorescence microscope.

16. Take another 1 ml sample and after the microcarriers settle remove the supernatant for measurement of the LDH activity. Incubate the cells/beads with 1 ml of TrypLE to detach the cells. After 10 min, pipette the sample vigorously to separate all cells from microcarriers. Add equal amount of Trypan Blue to count live and dead cells. This number can be used to calculate the cell seeding efficiency and viability.

17. The supernatant from the last step can be used for LDH activity determination according to the manufacturer's instruction and thus assess the level of cell lysis transpiring in the medium.

3.2.2 Microcarrier
Culture Using Matrigel

1. Equilibrate FACT III collagen-coated polystyrene microcarriers (0.5 g microcarriers for five million cells seeding) in PBS for 5 min and autoclave them. Allow the beads to cool to 4 °C.

2. Wash the microcarriers with cold DMEM/F12 and mix well with cold liquefied Matrigel (200 µl of Matrigel solution for 0.5 g of microcarriers).

3. Leave the microcarrier/Matrigel mix at room temperature for 1 h. Pipette up and down every 10 min to get all microcarriers suspended.

4. After 1 h, equilibrate the coated microcarriers in culture medium containing 10 µM ROCK inhibitor for 1 h before cell seeding. Beads at 0.5 g ($\sim$180 cm^2 surface area)/50 ml medium were used.

5. For initial seeding on beads, treat the stem cells on Matrigel-coated dishes with 10 µM ROCK inhibitor for 1 h and dissociate the colonies into single cells by incubating with Accutase for 10 min.

6. Transfer the dispersed hPSCs with microcarriers (cell-to-bead ratios as stated) to Petri dishes and place in 5 % CO_2 at 37 °C.

7. Gently shake the dish to better mix the microcarriers with cells every 15 min.

8. Subsequently, the cell-laden beads are transferred to spinner flasks (Corning) and the total medium (with Y-27632) volume is brought to 50 ml.

9. The agitation rate is set to 45 rpm.

10. After the first day, the medium is replaced with the same medium but without Y-27632.

11. Subsequent medium changes are performed at half-volume every day.

12. The cultures are maintained at 37 °C in 5 % CO_2. Cells on microcarriers are shown in Fig. 1b after staining with FDA.

13. In preparation for passaging cells between spinner flasks, Y-27632 is added to the culture at a final concentration of 10 µM for about 1 h prior to harvesting.

14. Collect cells on microcarriers from spinner flasks into 50 ml centrifuge tubes.

15. Wash the cells once with PBS: allow the microcarriers to settle and remove as much of the medium as possible. Add 10 ml PBS to the beads and resuspend them. Incubate the beads with Accutase for 10–15 min in the incubator and gently mix them occasionally.

16. Complete detachment from microcarriers as single cells can be verified by microscopy. Pass the cell/bead suspension through a 100-μm mesh strainer. The harvested cells can be characterized and/or inoculated onto fresh microcarriers repeating the steps described above.

3.3 Aggregate Culture of hPSCs in Spinner Flasks

1. Wash and autoclave the spinner flasks to be used.

2. Incubate five million cells in dishes with 10 μM ROCK inhibitor for 1 h.

3. After 1 h, dissociated the colonies into single cells by incubating them in Accutase for 10 min.

4. Centrifuge the dispersed hPSCs at $200 \times g$ for 5 min and remove the supernatant.

5. Resuspend the cell pellet in 2 ml of E8 medium supplemented with 10 μM ROCK inhibitor and 0.1 ml cell coating matrix (Life Technologies) for 10 min.

6. Transfer the cell suspension to ProCulture spinner flasks (Corning). The total medium volume supplemented with ROCK inhibitor is brought to 50 ml.

7. The agitation rate is set to 45 rpm.

8. After the first day, replace the medium with medium not containing Y-27632.

9. Subsequent medium changes are performed at half-volume every day.

10. The cultures are maintained at 37 °C in 5 % CO_2.

11. In preparation for passaging cells between spinner flasks, Y-27632 is added to the culture at a final concentration of 10 μM about 1 h prior to harvesting.

12. Then, cell aggregates are collected from spinner flasks into centrifuge tubes and incubated with Accutase for 10–15 min while gently mixing occasionally.

13. Images of aggregates as well as the growth profile, viability, and expression of pluripotency markers of a 6-day culture of H9 hESCs are shown (Fig. 1c–e).

3.4 Differentiation to Definitive Endoderm and Mesoderm Cells

1. After the microcarriers become confluent with cells (usually in 6 days of culture), differentiation can be initiated.

2. Before changing to differentiation medium, let the beads with cells settle for 10 min.

3. Remove as much culture medium as possible from the flask.

4. Add 20 ml of warm DMEM/F12 medium to the spinner flask and shake gently to resuspend the beads.

5. Let the beads settle again for another 10 min.

6. Remove as much medium as possible while being careful not to aspirate cells with beads.

7. Add 50 ml of fresh differentiation medium to the spinner flask and set the stirring rate to 45 rpm.

8. Change medium every day for 4 days by letting the beads settle, aspirating carefully most of the existing medium and replacing it with equal volume of fresh medium.

9. For definitive endoderm differentiation, the differentiation medium is composed of RPMI 1640 medium containing 100 ng/ml activin A and 1 % CTS-B27 (xeno-free).

10. For mesoderm differentiation, the differentiation medium is composed of RPMI 1640 medium containing 10 ng/ml activin A, 10 ng/ml BMP4, and no KSR (day 1 of differentiation), 0.2 % KSR (day 2), or 2 % KSR (days 3–4).

3.5 Characterization of Cultured Cells

3.5.1 Reverse Transcription-Quantitative PCR

All procedures must be done on RNase-free surfaces. Wipe the benchtop, and all relevant equipment with RNase Away solution (Thermo Scientific). Use RNase/DNase-free tubes and filtered pipet tips only. Use a fume hood for RNA extraction. Work with all samples on ice for the cDNA synthesis and qPCR steps.

1. Isolate RNA from the harvested cells ($1-10 \times 10^6$ cells) using 1 ml Trizol. Pellet the harvested cells by centrifuging at $200 \times g$ for 5 min. Add 1 ml Trizol to the cells. Mix the cell pellet with the Trizol using a 1-ml pipet tip, and triturate vigorously to remove any signs of the pellet. Allow the solution to incubate at room temperature for 5 min.

2. Add 200 µl chloroform to the mixture, and allow it to incubate at room temperature for 2–3 min. Shake the tube vigorously for 15 s.

3. Centrifuge at $12,000 \times g$ and 4 °C for 15 min. There will be three layers: a lower (pink) organic phase, a middle (white) phase, and a top (clear) aqueous phase. Using a 200-µl pipet tip, remove the aqueous phase and transfer it into a fresh tube taking care not to disturb the other phases.

4. Add 500 µl of isopropanol to the aqueous phase. Add 0.5 µl glycogen to the mixture to visualize better the RNA pellet.

Shake the tube vigorously for 10 s. Allow to incubate at room temperature for 10 min, and centrifuge at $12,000 \times g$ and $4\,^{\circ}\text{C}$ for 10 min. A white pellet will be visible at the bottom of the tube (if using a swinging bucket rotor; for a fixed-angle rotor, it will be visible at the outer lower edge of the tube).

5. Remove the isopropanol by upending the tube and add 75 % ethanol to the pellet. Vortex the pellet so that it floats in the ethanol (this is a washing step). Centrifuge at $7500 \times g$ and $4\,^{\circ}\text{C}$ for 5 min.

6. Upend the tube to remove the majority of the ethanol, and allow the rest of it to air-dry taking care not to let the pellet dry completely.

7. Resuspend the pellet in 30–200 µl nuclease-free water depending on the size of the pellet and measure the concentration using a UV spectrometer. Measure the 260 nm-to-280 nm absorbance ratio to ensure the purity of the RNA (ideally this should be above 1.8).

8. Aliquot 1 µg RNA in a nuclease-free PCR tube. Add 1 µl oligodT primers, and enough nuclease-free water to make the volume up to 5 µl.

9. Program the thermal cycler to hold the samples at 70 °C for 5 min and 4 °C for at least 10 min, and place the samples in the cycler.

10. Add 2.4 µl $MgCl_2$, 4 µl 5× reaction buffer, 1 µl dNTP mix, 6.4 µl nuclease-free water, 0.5 µl RNasin RNase inhibitor, and 1 µl reverse transcriptase to the samples to make a total volume of 20 µl.

11. Program the thermal cycler to hold the samples at 25 °C for 5 min to anneal the samples, 42 °C for 1 h to extend the chain, 70 °C for 15 min to inactivate the reverse transcriptase, and 4 °C to cool the samples and hold them until they can be used for qPCR (store samples at $-20\,^{\circ}\text{C}$ if not performing qPCR on them immediately). Add a melting curve cycle to check for nonspecific products. Place the samples in the cycler.

12. Place the 96-well qPCR plate on a plate holder. Each sample would contain 10 µl SYBR Green master mix, 8 µl nuclease-free water, 0.5 µl of the forward primer, 0.5 µl of the reverse primer (forward and reverse primers are specific for the genes the expression of which will be probed), 0.2 µl of the ROX dye (for Applied Biosystems qPCR platforms), and 1 µl of the cDNA. Prepare aliquots of the mixture for running triplicate samples. Each sample will be probed for *ACTB* (b-actin), *NANOG*, *POU5F1* (OCT4), and *SSEA4*.

13. Seal the plate using sealing film, and insert in the qPCR instrument.

3.5.2 Flow Cytometry

Use a no-primary antibody control (NPAC) or an isotype control (for conjugated primary antibodies).

1. Make a 1 % NDS solution for washes. Make a 3 % NDS solution for blocking (*see* **Note 5**).

2. Use one million of the harvested cells per sample, preparing samples to be tested for NANOG, OCT4, SSEA4, and TRA-1-60.

3. Add 1 ml of 4 % paraformaldehyde solution to each sample. Incubate at room temperature for 10 min tapping every 2 min to keep cells in suspension. Do not vortex or pipet.

4. Centrifuge the sample at $1500 \times g$ for 5–10 min. Remove the paraformaldehyde solution in the fume hood. Add 1 ml of 1 % NDS solution and resuspend the pellet by tapping and vortexing. Centrifuge the samples at $1500 \times g$ for 5–10 min to be able to collect all the cells. Remove the NDS solution.

5. Add 100 μl cytonin to each pellet and tap tubes several times to resuspend cells. Incubate at room temperature for 1 h. Tap occasionally to keep the cells in suspension.

6. Remove cytonin by first adding 1 ml of 1 % NDS to the samples and then spinning down at $1500 \times g$ for 5–10 min.

7. Block the samples by adding 600 μl of 3 % NDS. Keep cells in suspension during the wait by tapping them at regular intervals. Incubate for 1 h at room temperature. Spin down at $1500 \times g$ for 5–10 min.

8. Prepare primary antibody solutions in 1 % NDS at dilutions according to the manufacturers' instructions. Incubate for 40 min to an hour at room temperature. For an NPAC add 1 % NDS only. Keep the samples in suspension by tapping them every 10 min.

9. Wash each sample three times with 1 % NDS.

10. Make a solution of appropriate secondary antibodies in 1 % NDS. Add 100 μl secondary antibody to the samples and resuspend them. Incubate the samples at room temperature for 1 h in a dark place.

11. Wash three times with 1 % NDS.

12. Add 250–500 μl PBS to each pellet and resuspend the samples.

13. Examine the samples using a flow cytometer.

3.5.3 Immunohistochemistry

1. Coat microslides with vitronectin or Matrigel following the steps above.

2. Seed the harvested single cells onto the slides such that each sample has two wells: one experimental well and one NPAC well.

3. After 1 or 2 days of culture, add PBS to the cells. Aspirate the PBS and add 4 % paraformaldehyde solution to the cells in a fume hood. Incubate for 15–20 min.

4. Meanwhile, make a blocking/permeabilizing solution by adding 10 µl Triton X and 100 mg BSA to PBS to make a 0.1 % Triton-X and 1 % BSA solution.

5. Wash the cells three times with PBS. Incubate the cells for 5 min in PBS before aspirating it and adding fresh PBS. This constitutes one wash step.

6. Add the blocking/permeabilizing solution to the cells and incubate for 30 min.

7. Wash three times with PBS, incubating the cells for 5 min in PBS.

8. Prepare the primary antibody solution in 1–2 % serum (*see* **Note 6**) at dilutions recommended by the manufacturer. NANOG, OCT4, SSEA4, and TRA-1-60 are some of the markers tested.

9. Incubate the cells in primary antibody solution for 1 h at room temperature or overnight at 4 °C. Take care not to add the primary antibody solution to the wells with no primary control.

10. Wash the cells three times with PBS.

11. Prepare the secondary antibody solution by adding an appropriate amount of secondary antibody to 1–2 % serum solution.

12. Incubate the cells in the secondary antibody solution for 1 h at room temperature.

13. Wash the cells three times in DI water. Prepare a DAPI solution according to the manufacturer's instructions by adding an appropriate amount of DAPI to DI water.

14. Incubate in DAPI for 5 min.

15. Wash three times in DI water. Aspirate all the DI water, and remove the cover and gasket with the wells from the slide.

16. Add one to two drops of gel mounting solution to the slide. Carefully add the cover slip. Incubate for 1 h in the dark at room temperature.

17. Seal the cover slip using nail polish or other appropriate sealant.

3.5.4 Karyotyping

1. Cells harvested from spinner flask cultures are re-plated on T-75 flasks and allowed to grow until 70 % confluence.

2. Cells are treated with 30 ng/ml of KaryoMAX Colcemid Solution (Gibco) for 4 h at 37 °C.

3. Cells are then collected, transferred to 15-ml conical tubes, centrifuged for 5 min at 250 × g, and gently resuspended in cell hypotonic solution (CHS: 40 mM KCl, 20 mM HEPES, 0.5 mM EGTA, and 9 mM NaOH) for 1 h of incubation at 37 °C.

4. After centrifugation of the cell/CHS suspension at 250 × g, the supernatant is removed and the cells are fixed with 1:3 (v/v) acetic acid:methanol solution. Resuspend by pipetting up and down.

5. Spin at 200 × g for 5 min and remove supernatant. Repeat twice the fixing wash with methanol/acetate.

6. Resuspend the cells in 2 ml methanol/acetate. Apply one drop of the cell suspension to a glass slide. Keep the pipette at a considerable distance from the slide so that the drop can spread on the slide.

7. Allow the slide to dry.

8. Dip the slide in staining solution (Wright-Giemsa stain modified) or add the solution drop by drop to the slide until all the slide surface is covered. Transfer to PBS to wash and let the slide air-dry.

9. Alternatively, the samples can be stained with DAPI (instead of the Wright-Giemsa stain solution) and GelMount solution is applied before covering each slide with a cover slip [12].

10. Slides are viewed under a microscope and at least 20–25 metaphase spreads are examined per preparation. This analysis reveals only gross karyotypic abnormalities or polyploidy.

3.5.5 Differentiation of Harvested Cells into Mesoderm

1. Coat a 12- or 6-well plate with Matrigel or vitronectin.

2. To make 50 ml of the differentiation medium, add 500 µl RPMI vitamins, 500 µl nonessential amino acids, 500 µl sodium pyruvate, 500 µl penicillin/streptomycin, 1000 µl RPMI amino acids, 50 µl chemically defined lipids, 50 µl sodium selenite solution (made at a concentration of 5 µg/ml), and 12.5 µl holo-transferrin (made at a concentration of 2 mg/ml) to DMEM.

3. Use ROCK inhibitor-supplemented E8 medium to seed cells. Count the harvested cells using a cell counter or hemocytometer. Seed cells at a density of 4–8 × 10^4 cells/cm^2.

4. Expand the cells for 6 days to a density of around 0.5 × 10^6 cells/cm^2.

5. On day 0 of differentiation (day 6 of expansion) add 30 ng/ml BMP4 and 100 ng/ml Wnt3a to the differentiating medium. Incubate the cells for exactly 24 h.

6. Aspirate the medium and add fresh differentiation medium after 24 h (day 1) of differentiation.

7. On day 2 of differentiation, or exactly 24 h after medium change, add KY0211 at a ratio of 1:1000 to the differentiating cells. This medium can be used henceforth and replaced daily.

8. On day 5 of differentiation, test for KDR, PGDFRα, and C-KIT using flow cytometry and qPCR.

9. Mesodermal cells are KDR$^+$/PGDFRα$^+$ as well as C-KIT^{low} [13, 14].

3.5.6 Differentiation of Harvested Cells into Ectoderm

1. Coat a 6-cm dish with Matrigel or vitronectin. Seed the harvested cells onto the dish.

2. Make neural induction medium (NIM) and neural proliferation medium (NPM) [15]. To make 50 ml NIM, add 500 µl N2 supplement, 500 µl glutamax supplement, 500 µl B27 supplement, and 24.25 ml neurobasal medium to 24.25 ml of DMEM/F12. To make 50 ml NPM, add 250 µl N2 supplement, 250 µl B27 supplement, 500 µl glutamax supplement, 20 ng/ml FGF2, and 24.25 neurobasal medium to DEMD/F12.

3. When the cells have reached confluence, use collagenase to dissociate the cells into aggregates: incubate cells in 2 ml of collagenase (2 mg/ml) for 5 min. Scrape the colonies using a 10 ml glass pipet. Add an equal volume of NIM to the aggregates and centrifuge the suspension at $200 \times g$ for 5 min. Using a 1 ml pipette tip, seed the cells on a low-binding 6-cm dish.

4. Culture the cells in NIM for 6 days, changing medium every day, and then culture them in NPM for 7 days. Test for upregulation and translation of Nestin and βIII-tubulin using qPCR and immunohistochemistry.

4   Notes

1. Matrigel is routinely used for the culture of hPSCs in dishes or suspension. This substrate however is a chemically undefined matrix derived from mouse sarcoma cells. And while Matrigel provides cues for maintaining the pluripotency of stem cells [16], its use is limited to the culture of stem cells intended for in vitro use. Recombinant vitronectin has recently been shown to support cultured stem cells.

2. Although E8 is the medium described here, other media such as the mTeSR (Stemcell Technologies) can be utilized for the same procedures.

3. If cells do not attach properly the following can be considered: (1) Cells need to be in the exponential growth phase. If the colonies are too large, or the dishes are too confluent, the cells may not attach and grow properly. (2) If the cells are not gently seeded onto the beads, cell viability may be reduced by bead damage to the cells. (3) Shaking the dish at regular 15-min intervals during the seeding phase is important as cells may stick to the dish surface instead of the microcarriers.

4. Paraformaldehyde solution preparation along with pH measurement should be carried out in a fume hood wearing personal protective equipment such as lab coat, gloves, and goggles.

5. For flow cytometry and immunostaining, the blocking serum depends on the species in which the secondary antibody is synthesized. We use NDS with donkey-generated secondary antibodies.

6. Although a general protocol for immunostaining is given, following the manufacturers' instructions for the blocking and cell permeabilization steps is recommended. The given procedure can serve as a starting point for most antigen detection. However, in some cases such as flow cytometry detection of surface markers, live-cell staining can also be implemented.

Acknowledgements

Funding support has been provided by the National Institutes of Health (NHLBI, R01HL103709) to EST.

References

1. Murry CE, Keller G (2008) Differentiation of embryonic stem cells to clinically relevant populations: lessons from embryonic development. Cell 132(4):661–680

2. Schwartz SD, Regillo CD, Lam BL, Eliott D, Rosenfeld PJ, Gregori NZ, Hubschman J-P, Davis JL, Heilwell G, Spirn M, Maguire J, Gay R, Bateman J, Ostrick RM, Morris D, Vincent M, Anglade E, Del Priore LV, Lanza R (2015) Human embryonic stem cell-derived retinal pigment epithelium in patients with age-related macular degeneration and Stargardt's macular dystrophy: follow-up of two open-label phase 1/2 studies. Lancet 385 (9967):509–516

3. Keirstead HS, Nistor G, Bernal G, Totoiu M, Cloutier F, Sharp K, Steward O (2005) Human embryonic stem cell-derived oligodendrocyte progenitor cell transplants remyelinate and restore locomotion after spinal cord injury. J Neurosci 25(19):4694–4705

4. Jiang M, Lv L, Ji H, Yang X, Zhu W, Cai L, Gu X, Chai C, Huang S, Sun J, Dong Q (2011) Induction of pluripotent stem cells transplantation therapy for ischemic stroke. Mol Cell Biochem 354(1-2):67–75

5. Cho S-W, Moon S-H, Lee S-H, Kang S-W, Kim J, Lim JM, Kim H-S, Kim B-S, Chung H-M (2007) Improvement of postnatal neovascularization by human embryonic stem cell–derived endothelial-like cell transplantation in a mouse model of hindlimb ischemia. Circulation 116(21):2409–2419

6. Jing D, Parikh A, Canty JM Jr, Tzanakakis ES (2008) Stem cells for heart cell therapies. Tissue Eng Part B Rev 14(4):393–406

7. Kehoe DE, Jing D, Lock LT, Tzanakakis ES (2010) Scalable stirred-suspension bioreactor

culture of human pluripotent stem cells. Tissue Eng Part A 16(2):405–421

8. Krawetz R, Taiani JT, Liu S, Meng G, Li X, Kallos MS, Rancourt DE (2010) Large-scale expansion of pluripotent human embryonic stem cells in stirred-suspension bioreactors. Tissue Eng Part C Methods 16(4):573–582

9. Lock LT, Tzanakakis ES (2009) Expansion and differentiation of human embryonic stem cells to endoderm progeny in a microcarrier stirred-suspension culture. Tissue Eng Part A 15 (8):2051–2063

10. Bardy J, Chen AK, Lim YM, Wu S, Wei S, Weiping H, Chan K, Reuveny S, Oh SK (2013) Microcarrier suspension cultures for high-density expansion and differentiation of human pluripotent stem cells to neural progenitor cells. Tissue Eng Part C Methods 19 (2):166–180

11. Jing D, Parikh A, Tzanakakis ES (2010) Cardiac cell generation from encapsulated embryonic stem cells in static and scalable culture systems. Cell Transplant 19(11):1397–1412

12. Kehoe DE, Lock LT, Parikh A, Tzanakakis ES (2008) Propagation of embryonic stem cells in stirred suspension without serum. Biotechnol Prog 24(6):1342–1352

13. Yang L, Soonpaa MH, Adler ED, Roepke TK, Kattman SJ, Kennedy M, Henckaerts E, Bonham K, Abbott GW, Linden RM, Field LJ, Keller GM (2008) Human cardiovascular progenitor cells develop from a KDR+ embryonic-stem-cell-derived population. Nature 453 (7194):524–528

14. Kattman SJ, Witty AD, Gagliardi M, Dubois NC, Niapour M, Hotta A, Ellis J, Keller G (2011) Stage-specific optimization of activin/nodal and BMP signaling promotes cardiac differentiation of mouse and human pluripotent stem cell lines. Cell Stem Cell 8(2):228–240

15. Nat R, Nilbratt M, Narkilahti S, Winblad B, Hovatta O, Nordberg A (2007) Neurogenic neuroepithelial and radial glial cells generated from six human embryonic stem cell lines in serum-free suspension and adherent cultures. Glia 55(4):385–399

16. Hughes CS, Postovit LM, Lajoie GA (2010) Matrigel: a complex protein mixture required for optimal growth of cell culture. Proteomics 10(9):1886–1890

Methods in Molecular Biology (2016) 1502: 53–61
DOI 10.1007/7651_2015_311
© Springer Science+Business Media New York 2015
Published online: 20 January 2016

Expansion of Human Induced Pluripotent Stem Cells in Stirred Suspension Bioreactors

Walaa Almutawaa, Leili Rohani, and Derrick E. Rancourt

Abstract

Human induced pluripotent stem cells (hiPSCs) hold great promise as a cell source for therapeutic applications and regenerative medicine. Traditionally, hiPSCs are expanded in two-dimensional static culture as colonies in the presence or absence of feeder cells. However, this expansion procedure is associated with lack of reproducibility and low cell yields. To fulfill the large cell number demand for clinical use, robust large-scale production of these cells under defined conditions is needed. Herein, we describe a scalable, low-cost protocol for expanding hiPSCs as aggregates in a lab-scale bioreactor.

Keywords: Induced pluripotent stem cells, Bioreactor, Pluripotency, Energy metabolism

1 Introduction

Adult somatic cells can be reprogrammed into induced pluripotent stem cells (iPSCs) through the expression of exogenous transcription factors [1]. Human (h) iPSCs are an attractive candidates for use in cellular-based therapies, as they circumvent ethical and immunological barriers associated with human embryonic stem cell (hESC) [2]. hiPSCs have the capacity for self-renewal and can differentiate into any cell type within the body. iPSCs can drug screening, drug discovery, and toxicology assays. Also, hiPSCs offer a unique platform for in vitro human disease modeling. Patient-specific iPSCs and their differentiated derivatives can provide valuable information about disease pathogenesis [3, 4]. In fact, hiPSCs may revolutionize personalized medicine as it allows the creation of genetically customized cell lines to each patient and through the mode of responders versus non-responder "trial in a dish" models [5, 6].

One of the for front challenges that impedes the implementation of hiPSCs is its production at a relevant clinical number. Conventionally, hiPSCs are expanded in two-dimensional static culture [7]. They are typically grown under feeder or feeder-free conditions in the presence of basic fibroblast growth factor (bFGF) to maintain pluripotency [8, 9]. Most of these methods use two dimensional adherent culture, which present difficulties for large-scale and production. Other disadvantage with static culture systems are such as time consumption in feeding and passaging, culture-to-culture variability, and nonhomogeneous culture conditions. Consequently, these culture systems are not good candidates for clinical application.

Scale-up production of hiPSCs in relevant clinical quantities can be achieved with the use of stirred suspension bioreactors, which offer several advantages over conventional adherent culture systems. Suspension bioreactors facilitate the large-scale expansion of hiPSCs required for clinical studies with lower cost [10–12]. They provide more homogeneous culture environment and offer a flexible platform for various modes for culturing cells and/or derivative tissues as aggregates, on microcarriers, or on scaffolds [13]. Suspension bioreactors maintain allow for online monitoring and control of culture parameters [12] while maintaining cell pluripotency, which can be determined by marker analysis or functional tests such as differentiation potential or energy metabolism [14, 15], found in suspension bioreactors seems to induce pluripotency in cells grown in this vessel [16, 17]. Here, shear stress seems to cell pluripotency [18, 19]. Thus, suspension bioreactors is ideal for the development of standardized, fully controlled, and scalable culture processes to produce a clinically relevant quantity of hiPSCs. A detailed protocol for bioreactor expansion of hiPSCs is described in the following sections.

2 Materials

All medium and reagents should be cell culture-tested reagents and prepared in sterile conditions. Filtering non-sterile reagents through 0.2 mm filters is highly recommended. It is also strongly recommended to test each batch of hiPSC culture reagents to prevent adverse effects of batch-to-batch variation.

2.1 hiPSC Culture Reagents

1. hiPSC medium: mTeSR™ feeder-free medium (Stem Cell Technologies, Vancouver, Canada) is recommended. This medium comes in 500 mL bottles with separate base medium and growth factor components. 200 mL of mTeSR™ will be required for culture of hiPSCs in each 100 mL bioreactor.

2. Extracellular matrix: Matrigel™ ECM (BD Biosciences, Franklin Lakes, NJ) is recommended for static culture using mTeSR™ medium. Resuspend Matrigel™ and coat the culture dishes according to the manufacturer's instructions.

3. Dissociation enzymes: Accutase™ (Stem Cell Technologies) is recommended for single-cell dissociation of hiPSCs in both static and suspension culture systems. Other enzymes, such as

Collagenase and trypsin, are not recommended for dissociation of hiPSC aggregates derived in suspension culture system (*see* **Note 4**). Accutase™ should be stored at −20 °C. It is highly recommended to aliquot Accutase™ into smaller volumes since repeated freeze-thaw cycles or extended storage at 4 °C will reduce Accutase™ activity.

4. ROCK inhibitor: 10 μM Y-27632 (Stemgent, Cambridge, MA) treatments are recommended for single-cell dissociation steps.

2.2   Cell Counting and Viability Assessment

1. Trypan Blue (Bio-Rad, Hercules, CA) and the Bio-Rad cell counting system are used to calculate the total and viable cells in static and suspension culture. Alternatively, manual counting using a hemocytometer or any other cell counting and viability assessment method can be used.

2.3   Pluripotency Assessment

1. PCR: Reverse transcriptase (RT)- and quantitative (q)-PCR are conducted to determine the expression of pluripotency-related genes including Oct-4, Sox2, Klf4, Nanog, SSEA-3, SSEA-4, and REX-1.

2. Flow cytometry: Fixed, permeabilized, and blocked hiPSCs are incubated with fluorescently labeled antibodies against Oct-4, Nanog, SSEA-3, SSEA-4, and REX-1 (all BD, Biosciences) according to the manufacturer's instructions. The labeled cells are then analyzed using a flow cytometer (BD) registering at least 10,000 events.

3. Teratoma assay: One million dissociated hiPSCs are suspended in $1 \times$ PBS$^-$ (Invitrogen) and injected into the thigh muscle of an immunodeficient SCID mouse (Taconic Farms, Hudson, NY). Tumor cell masses are usually formed in 4–6 weeks. Sections from newly formed tissue are analyzed by regular histology procedures to determine the presence of three germ layers in obtained tissues.

2.4   Anaerobic Energy Metabolism Assessment

Manipulating glycolic over cellular respiratory pathways is a useful indicator of iPSC pluripotency. The growing condition of iPSCs in bioreactors supports glycolysis which is an ideal condition for pluripotency. Glycolytic energy metabolism can be confirmed through different tests; however, in the following we show the most common tests, which are applicable and easily adaptable to most of the laboratories.

1. PCR: RT- and q-PCR are conducted to determine and quantify the expression of mitochondrial biogenesis genes as indicators of mitochondrial numbers. The genes are ND1, ND5, and MT-CYB.

2. Flow Cytometry: Mitochondrial mass and level of reactive oxygen species (ROS) are measured by flow cytometry following treatment of hiPSCs with $2',7'$-dichlorodihydrofluorescein diacetate (H_2DCFDA) (Molecular Probes®, Life Technologies, Carlsbad, CA) and Mito-tracker green (Molecular Probes®, Life Technologies, Carlsbad, CA), respectively.

3. Bioluminescence Luciferase-based Assay: The level of intracellular ATP is measured using ATPLite bioluminescence luciferase-based assay (Perkin Elmer, Waltham, MA). hiPSCs are used as input cells and then luminescence is quantified with a luminometer (Berthold Technologies, Bad Wildbad, Germany), following the manufacturer's instruction. About 100,000 cells are used as input. The results are presented as nanomoles of ATP per cell.

2.5 Suspension Bioreactor Materials

1. Bioreactor vessel: 100 mL suspension bioreactors (NDS Technologies, Vineland, NJ) with magnetic impellers should be siliconized prior to use with Sigmacoat (Sigma) following the manufacturer's instructions. Siliconization of the vessel is important to prevent attachment of hiPSC aggregates to vessel's glass surface. This process should be repeated every 6 months or as required.

2. Magnetic stir plate: Magnetic stir plates with variable speed are required. There are many suppliers for stir plates in the market. Stir plates from VWR (VWR International, Radnor, PA) are used in our lab. Please ensure that stir plate can maintain a speed of 100 rpm, and can be housed in a cell culture incubator. If your unit is not designed to be housed at 37 °C, the temperature in the incubator may reach 90–100 °C effectively killing all mammalian cells.

3   Methods

3.1 Preparation of hiPSCs in Static Culture

1. hiPSCs can either be cultured under feeder-dependent or feeder-free conditions. Under feeder-dependent conditions either inactivated (*see* **Note 1**) human or murine fibroblasts can be used. Under feeder-free conditions Matrigel™ is recommended, used according to the manufacturer's instructions. When using mTeSR™, hiPSCs medium is privileged (*see* **Note 2**) for static culturing of hiPSC, as this medium will be used within the suspension bioreactor.

2. For single-cell passaging under static culture conditions, it is recommended to use Accutase™ and the ROCK inhibitor Y-27632. Normally, hiPSCs are treated with 10 μM Y-27632 1 h before dissociation and for 24 h after passaging (*see* **Note 3**).

3. One million viable hiPSCs are required to seed each 100 mL suspension bioreactor vessel. Therefore, it is recommended to have at least double that amount in static culture per bioreactor, since a number of cells will undergo apoptosis during single-cell dissociation even in the presence of Y-27632.

3.2   Transferring hiPSCs from Static to Suspension Culture

1. Add Y-27632 (10 μM) to the culture medium of static-cultured hiPSCs 1 h prior to single-cell dissociation (*see* **Note 4**).

2. Discard the culture medium and add a sufficient amount of Accutase™ to cover the whole dish/es. Incubate the hiPSCs at 37 °C, and check cell dissociation every 1–2 min (*see* **Note 5**).

3. Stop enzymatic reaction by adding fresh medium, and break up any cell clumps with gentle pipetting. Pellet the hiPSCs by centrifugation at $800 \times g$ for 5 min.

4. Discard the supernatant and resuspend the cells at a density of one million/milliliter in a sterile 15 mL conical tube containing fresh mTeSR™ medium with 10 μM Y-27632. Incubate the cells in suspension culture (in low-adherent tissue culture plates) for 1 h at 37 °C in 5 % CO_2 (*see* **Note 6**).

5. Prepare a 100 mL suspension bioreactor with 100 mL of fresh, pre-warmed (*see* **Note 7**) mTeSR™ medium with 10 μM Y-27632 and 0.1 nM Rapamycin.

6. Add one million hiPSCs to each 100 mL suspension bioreactor and place the bioreactors containing hiPSCs at 37 °C, 5 % CO_2, and 100 rpm (*see* **Note 8**).

3.3   Culturing hiPSCs in Suspension

1. It is highly recommended to change the medium 24 h after initial seeding (*see* **Note 9**).

2. Collect all 100 mL of medium with hiPSCs in two sterile 50 mL conical. Spin the conicals at $800 \times g$ for 5 min, and carefully remove as much of the supernatant as possible with aspiration. Do not pour off the old medium as the small hiPSC aggregates will not adhere to the bottom of the conical and will be lost. We recommend leaving the last 1–5 mL of old medium per conical so as not to accidently remove the hiPSCs.

3. Prepare the bioreactor with 100 mL of new pre-warmed mTeSR™ medium and 0.1 nM Rapamycin. It is not recommended to add Y-27632.

4. Place the bioreactor vessel at 37 °C, 5 % CO_2, and 100 rpm. The medium usually does not need replacing until day 6 of culture; however, it can be replaced as often as desired.

3.4   Passaging of hiPSCs in Suspension

1. On day 6 of bioreactor culture, medium and cells are collected from the bioreactor into two 50 mL sterile conical tubes. Centrifuge 50 mL tubes at $800 \times g$ for 5 min. Discard

(without pouring) old medium leaving 1–2 mL. Add 10 μM Y-27632 to the medium and cells and incubate at 37 °C, in 5 % CO_2 for 1 h.

2. Transfer hiPSCs to a 15 mL sterile conical tube and spin down at $800 \times g$ for 5 min. Discard as much of the old medium as possible, and add 1 mL of Accutase™. Incubate the cells at 37 °C, 5 % CO_2 checking occasionally under a dissecting microscope to observe dissociation of aggregates. Following aggregate dissociation into single cells, break up any small clumps by gentle agitation.

3. Once the majority of the aggregates have been dissociated into single cells, break up any small clumps by gentle pipetting.

4. Repeat steps 3–6 under Section 4.2, and steps 1–4 under Section 4.3.

3.5 Expansion Rate Calculation

It is recommended to count hiPSCs and assess the viability at each day of bioreactor culture by the following steps:

1. Take only 5 mL sample from the bioreactor and repeat steps 1–3 of Section 4.4 (minding only to take 5 mL and not the entire volume).

2. Add new 5 mL of fresh medium to the bioreactor.

3. Using Trypan Blue and an automated cell counter or hemocytometer, record the percent viability and total cell count.

3.6 Assessment of Pluripotency and Anaerobic Energy Metabolism

At the conclusion of your experiment, it is recommended to analyze the pluripotency of the hiPSCs. It would be beneficial to check pluripotency features of hiPSCs every three to five passages following their culture in bioreactors. We usually assess the pluripotent state of hiPSCs through FACS, PCR, and teratoma analysis (*see* **Note 10**). Analysis of anaerobic energy metabolism as an indication of proliferating hiPSCs in bioreactors is also recommended. Normally, we evaluate this through PCR, flow cytometry, and bioluminescence luciferase-based assay (*see* **Note 11**).

4   Notes

1. Many different methods have been reported for the static culture of hiPSCs previously [20–22]. However, only culturing methods that use mTeSR™ are suitable for suspension culture. Therefore, hiPSCs should be adapted to mTeSR™ before transferring them to suspension culture.

2. The method which is described here to grow hiPSCs in suspension utilizes mTeSR™ hiPSCs medium. Other culture

mediums may also work but culturing conditions and other growth factors may need optimization.

3. Y-27632 and other ROCK pathway inhibitors have been shown to allow hiPSCs dissociation into single cells without massive cell death [23, 24]. However, for the successful generation of hiPSC aggregates in suspension only Rho/ROCK pathway inhibitors that target ROCK or upstream of ROCK are effective.

4. It is essential to use single-cell inoculation into bioreactors, as it helps to keep the undifferentiated state of hiPSC aggregates. If partial collagenase dissociation is utilized, the risk of generating EBs is greatly increased. However, chromosomal changes can occur following prolonged enzymatic single-cell passaging of the iPSCs [25]. Hence, recently the new enzyme-free EDTA-based reagent called Versene (Versene® (EDTA) 0.02 %, Lonza, Basel, Switzerland) has been offered to the market, allowing single cell dissociation of stem cells without affecting their chromosomal status. Although we have not used Versene for single-cell dissociation of hiPSCs, it may be a useful alternative for single-cell passaging of bioreactor-cultured hiPSCs.

5. Normally, in static culture hiPSC colonies are tightly packed with well-defined borders. With enzymatic dissociation, individual cells will be identifiable within the colony. It is recommended not to allow single-cell dissociation to proceed until all cells are released into the medium, as this is unnecessarily may damage hiPSC. Instead, wait until most of the cells are identifiable within the hiPSC colonies, stop the reaction with fresh medium, and break up any leftover cell clumps with gentle mechanical dissociation.

6. The 1-h incubation time with Y-27632 promotes the formation of small "seed" aggregates (1–5 cells) and thus enhances the efficiency of aggregation and expansion of hiPSCs following transferring to the stirred bioreactor.

7. Pre-warm all hiPSC culture medium to 37 °C to avoid shocking the cells.

8. To calculate rpm of the bioreactor, do not rely on the digital/analog settings on your stir plate. Each bioreactor design will behave differently when placed on the magnetic stir plate. Verify your rpms by counting the number of impeller rotations in 10s and multiply by 6 to get your revolutions per minute. One hundred rotations per minute is optimal for hiPSC aggregate formation and expansion, if significantly slower speeds are used (60–80 rpm), aggregates will grow quickly and may develop either differentiated or necrotic centers. However, at higher speeds (120+) cells do not effectively aggregate: any resultant small aggregates will normally die.

9. If the Y-27632 is not removed from the bioreactor after 24 h, the aggregates will still remain viable and undifferentiated; however, the expansion rate will be significantly negatively impacted. Furthermore, the prolonged maintenance of Y-27632 in the medium might have negative effect on differentiation potential of hiPSCs for downstream studies.

10. The following reference describes commonly used methods to analyze pluripotency of hiPSCs [26–28].

11. The following references describe anaerobic energy metabolism of hiPSCs and some of the tests discussed above [29–31].

References

1. Takahashi K, Tanabe K, Ohnuki M, Narita M, Ichisaka T, Tomoda K, Yamanaka S (2007) Induction of pluripotent stem cells from adult human fibroblasts by defined factors. Cell 131:861–872

2. Robinton DA, Daley GQ (2012) The promise of induced pluripotent stem cells in research and therapy. Nature 481:295–305

3. Bellin M, Marchetto MC, Gage FH, Mummery CL (2012) Induced pluripotent stem cells: the new patient? Nat Rev Mol Cell Biol 13:713–726

4. Grskovic M, Javaherian A, Strulovici B, Daley GQ (2011) Induced pluripotent stem cells—opportunities for disease modelling and drug discovery. Nat Rev Drug Discov 10:915–929

5. Hanna J, Wernig M, Markoulaki S, Sun CW, Meissner A, Cassady JP, Beard C, Brambrink T, Wu LC, Townes TM, Jaenisch R (2007) Treatment of sickle cell anemia mouse model with iPS cells generated from autologous skin. Science 318:1920–1923

6. Moretti A, Bellin M, Welling A, Jung CB, Lam JT, Bott-Flügel L, Dorn T, Goedel A, Höhnke C, Hofmann F, Seyfarth M, Sinnecker D, Schömig A, Laugwitz KL (2010) Patient-specific induced pluripotent stem-cell models for long-QT syndrome. N Engl J Med 363:1397–1409

7. Kirouac DC, Zandstra PW (2008) The systematic production of cells for cell therapies. Cell Stem Cell 3:369–381

8. Yu J, Vodyanik MA, Smuga-Otto K, Antosiewicz-Bourget J, Frane JL, Tian S, Nie J, Jonsdottir GA, Ruotti V, Stewart R, Slukvin II, Thomson JA (2007) Induced pluripotent stem cell lines derived from human somatic cells. Science 318:1917–1920

9. Chen G, Gulbranson DR, Hou Z, Bolin JM, Ruotti V, Probasco MD, Smuga-Otto K, Howden SE, Diol NR, Propson NE, Wagner R, Lee GO, Antosiewicz-Bourget J, Teng JM, Thomson JA (2011) Chemically defined conditions for human iPSC derivation and culture. Nat Methods 8:424–429

10. Krawetz R, Taiani JT, Liu S, Meng G, Li X, Kallos MS, Rancourt DE (2010) Large-scale expansion of pluripotent human embryonic stem cells in stirred-suspension bioreactors. Tissue Eng Part C Methods 16:573–582

11. Kehoe DE, Jing D, Lock LT, Tzanakakis ES (2010) Scalable stirred-suspension bioreactor culture of human pluripotent stem cells. Tissue Eng Part A 16:405–421

12. dos Santos FF, Andrade PZ, da Silva CL, Cabral JM (2013) Bioreactor design for clinical-grade expansion of stem cells. Biotechnol J 8:644–654

13. Badenes SM, Fernandes TG, Rodrigues CA, Diogo MM, Cabral JM (2015) Scalable expansion of human-induced pluripotent stem cells in xeno-free microcarriers. Methods Mol Biol 1283:23–29

14. Hunt MM, Meng G, Rancourt DE, Gates ID, Kallos MS (2014) Factorial experimental design for the culture of human embryonic stem cells as aggregates in stirred suspension bioreactors reveals the potential for interaction effects between bioprocess parameters. Tissue Eng Part C Methods 20:76–89

15. Day B, Rancourt DE (2013) Metabolic status of pluripotent cells and exploitation for growth in stirred suspension bioreactors. Biotechnol Genet Eng Rev 29:24–30

16. Shafa M, Day B, Yamashita A, Meng G, Liu S, Krawetz R, Rancourt DE (2012) Derivation of iPSCs in stirred suspension bioreactors. Nat Methods 9:465–466

17. Amit M, Chebath J, Margulets V, Laevsky I, Miropolsky Y, Shariki K, Peri M, Blais I, Slutsky G, Revel M, Itskovitz-Eldor J (2010) Suspension culture of undifferentiated human

embryonic and induced pluripotent stem cells. Stem Cell Rev 6:248–259

18. Fridley KM, Kinney MA, McDevitt TC (2012) Hydrodynamic modulation of pluripotent stem cells. Stem Cell Res Ther 3:45

19. Gareau T, Lara GG, Shepherd RD, Krawetz R, Rancourt DE, Rinker KD, Kallos MS (2014) Shear stress influences the pluripotency of murine embryonic stem cells in stirred suspension bioreactors. J Tissue Eng Regen Med 8:268–278

20. Ludwig TE, Bergendahl V, Levenstein ME, Yu J, Probasco MD, Thomson JA (2006) Feeder-independent culture of human embryonic stem cells. Nat Methods 3:637–646

21. Meng G, Liu S, Krawetz R, Chan M, Chernos J, Rancourt DE (2008) A novel method for generating xeno-free human feeder cells for human ES cell culture. Stem Cells Dev 17:413–422

22. Meng G, Liu S, Li X, Krawetz R, Rancourt DE (2010) Extra-cellular matrix isolated from foreskin fibroblasts supports long term xeno-free human embryonic stem cell culture. Stem Cells Dev 19:547–556

23. Watanabe K, Ueno M, Kamiya D, Nishiyama A, Matsumura M, Wataya T, Takahashi JB, Nishikawa S, Muguruma K, Sasai Y (2007) A ROCK inhibitor permits survival of dissociated human embryonic stem cells. Nat Biotechnol 25:681–686

24. Krawetz RJ, Li X, Rancourt DE (2009) Human embryonic stem cells: caught between a ROCK inhibitor and a hard place. Bioessays 31:336–343

25. Heng BC, Heinimann K, Miny P, Iezzi G, Glatz K, Scherberich A, Zulewski H, Fussenegger M (2013) mRNA transfection-based, feeder-free, induced pluripotent stem cells derived from adipose tissue of a 50-year-old patient. Metab Eng 18:9–24

26. O'Connor MD, Kardel MD, Eaves CJ (2011) Functional assays for human embryonic stem cell pluripotency. Methods Mol Biol 690:67–80

27. Meisner LF, Johnson JA (2008) Protocols for cytogenetic studies of human embryonic stem cells. Methods 45:133–141

28. Emre N, Vidal JG, Elia J, O'Connor ED, Paramban RI, Hefferan MP, Navarro R, Goldberg DS, Varki NM, Marsala M, Carson CT (2010) The ROCK inhibitor Y-27632 improves recovery of human embryonic stem cells after fluorescence-activated cell sorting with multiple cell surface markers. PLoS One 5, e12148

29. Prigione A, Fauler B, Lurz R, Lehrach H, Adjaye J (2010) The senescence-related mitochondrial/oxidative stress pathway is repressed in human induced pluripotent stem cells. Stem Cells 28:721–733

30. Varum S, Rodrigues AS, Moura MB, Momcilovic O, Easley CA, Ramalho-Santos J, Van Houten B, Schatten G (2011) Energy metabolism in human pluripotent stem cells and their differentiated counterparts. PLoS One 6, e20914

31. Armstrong L, Tilgner K, Saretzki G, Atkinson SP, Stojkovic M, Moreno R, Przyborski S, Lako M (2010) Human induced pluripotent stem cell lines show stress defense mechanisms and mitochondrial regulation similar to those of human embryonic stem cells. Stem Cells 28:661–673

Methods in Molecular Biology (2016) 1502: 63–75
DOI 10.1007/7651_2016_354
© Springer Science+Business Media New York 2016
Published online: 27 April 2016

Uniform Embryoid Body Production and Enhanced Mesendoderm Differentiation with Murine Embryonic Stem Cells in a Rotary Suspension Bioreactor

Xiaohua Lei, Zhili Deng, and Enkui Duan

Abstract

Embryonic stem cells (ESCs) are capable of differentiating into almost all cell types in vitro and hold great promise for drug screening, developmental studies and have a huge potential in many therapeutic areas. ESCs can aggregate to form embryoid body (EB) in static suspension culture by spontaneous differentiation, which resembles an intact embryo; while static suspension culture cannot prevent agglomeration of cells and offers little control over the size and shape of EBs, it results in aggregation of EBs into large, irregular masses, which prejudice the efficiency of differentiation of cells. Recently, bioreactor-based platforms have been shown to not only offer a beneficial effect on increasing diffusion of nutrients and oxygen which promotes cell viability and proliferation but also display local biomechanical properties (e.g., low fluid shear stresses and hydrodynamic force) in tissue development and organogenesis. This chapter describes a protocol for using a rotary suspension bioreactor to produce embryoid bodies and process the differentiation of mouse embryonic stem cells (mESCs), and to assess the efficiency of EB differentiation in the bioreactor by real-time PCR and immunostaining.

Keywords: Mouse embryonic stem cell, Rotary cell culture system, EB formation, Differentiation

1 Introduction

Embryonic stem cells (ESCs) provide an elegant model for studying cell fate determination and are promising cell sources for biomedical applications including regenerative medicine, cell therapy, and drug discovery [1, 2]. Differentiated ESCs possess a potentially unlimited source of transplantable cells. Thus, it is a prerequisite and very important for achieving clinical application to acquire specific types of differentiated cells with high purity in highly efficient and cost-effective ways. Though the commonly used embryoid body (EB) formation method can generate 3D cell aggregates in static suspension condition, which can be achieved by spontaneous differentiation, static suspension culture offers little control over the size and shape of EBs and often results in agglomeration of EBs into large, irregular masses, which will affect the efficiency of the differentiation of cells [3, 4]. Unlike traditional static culture methods, bioreactor systems have the ability to achieve large scale and can be integrated with chemical induced cell differentiation, both of which are critical for potential clinical applications [5–7].

Rotary cell culture system (RCCS) was originally designed by National Aeronautics and Space Administration (NASA) as an effective tool to simulate microgravity in space for cell growth. This 3D dynamic culture system has been recommended by researchers, which offers distinct advantages over classic static cultures in large-scale mammalian cell cultures, such as prevention of sedimentations, increased mass transfer and diffusion of oxygen [8]. In addition, the RCCS creates a low fluid shear and hydrodynamic force, which might strengthen tissue development and organogenesis [9, 10]. Therefore, RCCS has been used for expansion and differentiation of various stem cell types. Li et al. reported that human mesenchymal stem cells (MSCs) could be expanded about a 29-fold after 8 days of culture in RCCS bioreactor [11]. In keeping with these reports, we recently reported that RCCS can stimulate the proliferation of human epidermal stem cells and subsequent formation of 3D structure [12]. Moreover, we also demonstrated that RCCS can increase the efficiency of EB formation and augment mesendoderm differentiation of mouse embryonic stem cells (mESCs) by modification of Wnt/β-catenin signaling [13]. In the following section, we describe a protocol for using the rotary cell culture system to produce embryoid bodies and process differentiation of mouse embryonic stem cells (mESCs). We analyze the efficiency of EB differentiation in the rotary suspension bioreactor by real-time PCR and immunohistochemistry.

2 Materials

2.1 Cell and Cell Lines

1. Inactivated MEFs: Mouse embryonic fibroblasts (MEFs) are treated with mitomycin C and stored in liquid nitrogen (2×10^6 cells/vial).

2. ORG1: a mouse embryonic stem cell line that expresses GFP under promoter Oct4 [14].

2.2 Medium and Supplements for Cell Culture

1. PBS: Ca^{2+}, Mg^{2+} free, PBS (−) (Cat# 14190-144, Gibco), store at 4 °C.

2. Medium for inactivated MEF: High-glucose Dulbecco's modified Eagle's medium with GlutaMAX 1× (DMEM-GlutaMAX, Cat# 11960-044, Invitrogen), 10 % heat-inactivated fetal bovine serum (FBS, Cat# 11960-044, Gibco), 1× penicillin–streptomycin (Cat# 10378-016, Gibco), store at 4 °C.

3. Medium for mESC growth: DMEM-GlutaMAX, 15 % heat-inactivated FBS (Cat# 11960-044, Gibco) (see Note 1), 1 %

nonessential amino acid (NEAA, Cat# 11140-050, Gibco), 1×
penicillin–streptomycin (Cat# 10378-016, Gibco), 0.1 mM
beta-mercaptoethanol (Cat# 21985-023, Gibco), and
1000 U/ml recombinant LIF (Cat# ESG1107; Millipore).
Store at 4 °C.

4. EB induction and differentiation medium: DMEM-GlutaMAX
   with 20 % heat-inactivated FBS (Cat# 11960-044, Gibco), 1 %
   NEAA (Cat# 11140-050, Gibco), 0.1 mM 2-mercaptoethanol
   (Cat# 21985-023, Gibco). Stable for 2 weeks at 4 °C.

5. Solution of gelatin (0.2 %): gelatin from porcine skin
   (Cat#G1890 Sigma) is dissolved at 0.2 % solution in water,
   autoclaved and stored at 4 °C.

6. Collagenase IV solution: Collagenase IV (Cat# 17104-019,
   Gibco) power is dissolved at 1 mg/ml in PBS and stored at
   4 °C.

7. 0.05 % solution of trypsin–EDTA (Cat# 25300-054, Gibco),
   stored at 4 °C.

*2.3 Bioreactor
Equipment*

Rotary Cell Culture System (RCCS-4D, SYNTHECON), the
RCCS is composed of the following three parts.

1. Culture vessel (10 ml), sterilized clear plastic circular units with
   a half-inch diameter fill port (Fig. 1a) and two sampling or
   injection syringe ports (Fig. 1b) (*see* **Note 2**).

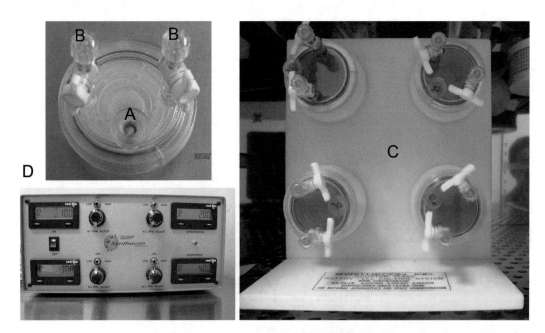

Fig. 1 Rotary bioreactor used in the described methodology. (**a, b**) 10 ml disposable vessel, including fill port
(**a**), and two syringe ports (**b**). (**c**) Rotator base, aims to support and rotate the culture vessel. (**d**) Power supply,
containing the electronic motor speed controls

2. Rotator Base, aims to support and rotate the culture vessel (Fig. 1c).

3. Power supply, containing the electronic motor speed controls. This equipment is used to adjust vessel rotation speed (Fig. 1d).

2.4 Real-Time PCR

1. TRIzol® Reagent (Cat# 15596018, Invitrogen, USA).

2. RQ1 RNase-free DNase I (Cat# M6101, Promega, USA).

3. M-MuLV reverse transcriptase (Cat# M0253L, New England Biolabs, China).

4. GoTaq® qPCR Master Mix (SYBR Green) (Cat# A6001, Promega, USA).

5. LightCycler® 480 II System thermocycler platform (Roche, USA).

2.5 Immunostaining

1. Fixative buffer: solubilization of 4 g paraformaldehyde (PFA, Cat# P-6148, Sigma) powder in 100 ml PBS (*see* **Note 3**). Store at −20 °C.

2. 1.5 % aqueous agar solution: 1.5 g agar power (Cat# A1296, Sigma) mixing with 100 ml water, heating to dissolve and at approx 60 °C.

3. Permeabilization buffer: 0.2 % Triton X-100 (Cat# 0694, Amresco, USA) in PBS. Store at −20 °C.

4. Blocking solution: 5 % bovine serum albumin (BSA, Cat# A1933, Sigma) in PBS or 5 % normal rabbit serum (Cat# 10510, Invitrogen) in PBS. Store at −20 °C.

5. Primary antibody: goat anti-brachyury polyclonal antibody (N-19) (Cat# sc-17743, Santa Cruz), rabbit anti-Gata6 monoclonal antibody (clone D61E4) (Cat# 5851S, Cell Signaling Technology), and rabbit anti-Sox1 polyclonal antibody (Cat# 4194S, Cell Signaling Technology).

6. Secondary antibodies: goat anti-rabbit TRITC-conjugated secondary antibodies (ZSGB BIO, Cat# ZF-0316 China) and rabbit anti-goat TRITC-conjugated secondary antibodies (ZSGB BIO, Cat# ZF-0317 China).

7. Nuclear staining solution: 0.1–1 µg/ml of Hoechst (bisBenzimide H33342 trihydrochloride) (Cat# B2261, Sigma) in PBS.

8. Nikon Ti microscope (Nikon, Japan).

3 Methods

3.1 Cultivation of mESCs on Inactivated MEF

1. Add 3 ml 0.2 % gelatin solution to 60 mm tissue culture dish. Incubate for 1 h at 37 °C incubator.

2. Remove the gelatin solution and wash twice with PBS for inactivated MEF seeding.

3. Thaw a vial of frozen inactivated MEFs (treated with mitomycin C and stored in liquid nitrogen, 2×10^6 cells/vial) quickly in a 37 °C water bath. Carefully transfer the cells into a 15 ml centrifuge tube. Add 10 ml of warm MEF medium, while gently shaking the tube.

4. Centrifuge at $200 \times g$ for 5 min, aspirate supernatant, and resuspend the pellet in 4 ml of MEF medium.

5. Seed onto the 0.2 % gelatin-coated 60 mm dish in MEF medium and place in a 37 °C, 5 % CO_2 incubator for 12–24 h.

6. Remove the MEF medium and wash cells once with sterile PBS (3 ml/dish) before plating of mESCs.

7. Thaw mESCs from liquid nitrogen by placing cryovials in a 37 °C water bath, gently move the contents from cryovials to 15 ml centrifuge tubes, add 6 ml prewarmed mESC growth medium, and centrifuge at $200 \times g$ for 5 min.

8. Remove supernatant, gently resuspend cells in 4 ml mESC growth medium, transfer them into the 60 mm dish coated with inactivated MEFs, place in a 37 °C incubator and shake the plate in quick side to side, forward to back motions to distribute the cells uniformly. Culture the cells at 37 °C incubator, with 5 % CO_2 and 95 % humidity.

9. Perform daily medium changes. Check for undifferentiated colonies (Fig. 2) that are ready for subculture (dense centered) or EB formation approximately 3–4 days after thawing.

3.2 RCCS Preparation Before Cell Culture Initiation

1. Unscrew the fill port and place fill port cap on a sterile 100 mm dish.

2. Fill the vessel by pipetting the 5 ml EB induction and differentiation medium through the open fill port.

3. Close the fill port with the cap and place the vessel to incubate for 10–30 min in the rinsing solution at 37 °C incubator and prepare for cell inoculation and cultivation.

4. Attach the multicolored ribbon cable of the power supply and the rotator base.

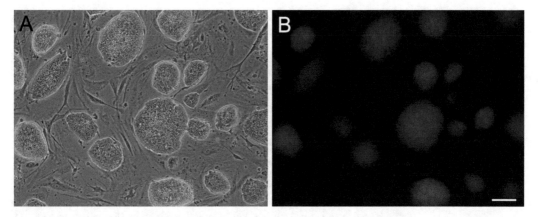

Fig. 2 Mouse embryonic stem (mES) cell colony appearance on MEF. When single cell suspensions were plated on MEF feeder, colonies showed up and remained undifferentiated. (**a**) Phase contrast of mES colony after 3 day cultured on MEF. (**b**) Oct4-GFP reporter fluorescence of mES colony after cultured on MEF. Bars: 50 μm

3.3 Single Cell Suspension Preparation for EB Induction in RCCS

1. Prior to EB induction, prepare single cell suspension from mESC colonies that have been cultured on inactivated MEFs as described above (*see* Subheading 3.1, **steps 8** and **9**).

2. Aspirate the mESC growth medium and rinse the cells once gently with 3 ml PBS.

3. Then, remove the PBS and add 2 ml solution of Collagenase IV (1 mg/ml, Invitrogen) to the 60 mm dish and incubate at room temperature for 2–3 min.

4. Add 2 ml EB induction and differentiation medium to the 60 mm dish and flush the dish gently to make mESCs detach as colonies.

5. Transfer dissociated mESC colonies to 15 ml centrifuge tubes, add 4 ml prewarmed EB induction and differentiation medium, and centrifuge at $50 \times g$ for 3 min.

6. Remove the supernatant and add 2 ml solution of trypsin–EDTA (0.05 %) to the 15 ml centrifuge tube, and incubate at 37 °C for 2–3 min.

7. Add 4 ml EB induction and differentiation medium to the tube, and gently pipette the cell suspensions 5–10 times to ensure any remaining clumps are fully dissociated.

8. Centrifuge the cells at $200 \times g$ for 5 min at room temperature (15–25 °C).

9. Aspirate the supernatant and resuspend the pellet in 10 ml of EB induction and differentiation medium.

10. Transfer the cell suspension to a 100 mm dish (Cat# 430166 Corning).

11. Incubate the cells at 37 °C with 5 % CO_2 and 95 % humidity for 30–40 min (*see* **Note 4**).

12. Harvest the non-adherent mESCs by drawing up the liquid from the dish. And transfer the suspension with non-adherent mESCs to a 15 ml centrifuge tube.

13. Centrifuge the harvested mESCs at $200 \times g$ for 5 min at room temperature (15–25 °C).

14. Remove and discard the supernatant of 15 ml centrifuge tube. Resuspend the cell pellet in 2 ml of EB induction medium.

15. Count cell number using the Scepter 2.0 Handheld Automated Cell Counter (Cat# PHCC20040 EMD Millipore) and Scepter with 40 μm Scepter Sensors according to the product instruction. For more information, please refer to the Product Information Sheet for the Scepter cell counter available on www.merckmillipore.com.

3.4 EB Induction and Differentiation in RCCS Bioreactor

1. Adjust the desired cell concentration suspension at $1–2 \times 10^6$/ml with EB induction medium.

2. Remove the rinsing solution in 10 ml vessel that is prepared as described above (*see* Subheading 3.2).

3. Open the two syringe ports and the fill port. Transfer 1 ml cell suspension into the 10 ml vessel through the fill port.

4. Add 6–8 ml EB induction medium until the vessel is nearly full and close the fill port with cap.

5. To remove the bubbles in vessel, place two syringes (5 ml) on each syringe port, one is a media-filled syringe which has 2–3 ml of EB induction medium and the other is an empty syringe.

6. Inject the medium in the media-filled syringe through the syringe port and make sure the bubbles are removed into the empty syringe.

7. Close the syringe valves and remove the syringes. Place the syringe caps on the syringe ports and close it (*see* **Note 5**).

8. Attach the 10 ml cell suspension-filled vessel onto the rotator base by slowly turning in clockwise direction.

9. Place the vessel and rotator base into the cell culture incubator at 37 °C with 5 % CO_2 and 95 % humidity.

10. Turn on the power supply and set the initial rotation speed to 15–20 rpm (*see* **Note 6**).

11. After 24–48 h, ES cells are allowed to aggregate spontaneously in RCCS bioreactor and EBs are formed (Fig. 3a, b). Media change should be performed every 2 days.

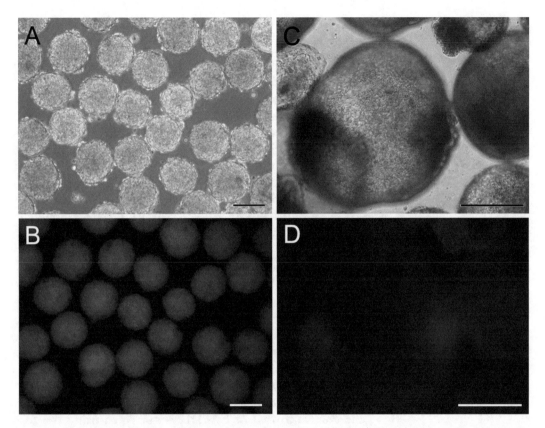

Fig. 3 Mouse embryoid body (mEB) appearance in rotary suspension culture. When mES cells are dissociated into single cell suspension and plated on RCCS bioreactor, they will form aggregates or embryoid bodies. These mEBs should appear homogeneous and as rounded spheres and increasingly cystic structures. (**a, b**) Phase contrast (**a**) and Oct4-GFP reporter fluorescence (**b**) of mEB after 24–48 h cultured in RCCS. (**c, d**) Phase contrast (**c**) and Oct4-GFP reporter fluorescence (**d**) of mEB cultured in RCCS for 8 days. Bars: 100 μm

12. For the medium change, turn off power and remove the vessel from the rotator base and take it to a biological hood. Let the cell aggregates settle to the bottom in static.

13. Open the valves and place two syringes (10 ml) on each syringe port; one is media-filled syringe which has 5–10 ml of EB induction medium and the other is empty syringe

14. Very carefully use the media-filled syringe to withdraw medium from the vessel as described above (*see* Subheading 3.2, **steps 6–8**). Usually, about 1/2 of the conditioned medium is left in the vessel and 1/2 of fresh medium will be added into the vessel (*see* **Note 7**).

15. Attach the vessel to the rotator base again and place them in the CO_2 incubator.

16. For 8 days of culture, EBs in RCCS exhibit a cystic structure (Fig. 3c, d).

3.5 Analysis of EBs Formed in RCCS: Real-Time PCR Detection

1. During EB differentiation, the EBs are collected at different time points.

2. For EBs collection, turn off power and remove the vessel from the rotator base and transfer the EBs from RCCS to a 15 ml centrifuge tube.

3. Perform the total RNA extraction, purification, reverse transcription, and real-time PCR according to the Product Information Sheet [15]. Briefly, EBs in different days of culture are collected by centrifugation, and the total RNA is purified with TRIzol reagent (Cat# 15596018 Life Technologies) (*see* **Notes 8** and **9**). The genomic DNA is eliminated using RQ1 RNase-free DNase I (Cat# M6101 Promega), and the cDNA is then synthesized with M-MuLV reverse transcriptase (Cat# M0253S, New England Biolabs). qPCR is performed using GoTaq® qPCR Master Mix (SYBR Green) (Cat# A6001 Promega) in the LightCycler® 480 II System thermocycler platform (Roche, USA).

4. The real-time PCR procedure is 94 °C for 2 min followed by 40 cycles of 95 °C for 15 s, 55 °C for 15 s, and 68 °C for 25 s. Forward and reverse primers used in qPCR reactions (*see* Table 1). Primers are verified by melting curve examination, and their amplification efficiencies are calculated using a relative standard curve method [16, 17]. mRNA expression of each gene relative to that of GAPDH is calculated using the ΔCt method with efficiency correction (Fig. 4).

Table 1
Primers used for qPCR experiments

Gene	Forward primer (5′-3′)	Reverse primer (5′-3′)
Gapdh	CATGTTCCAGTATGACTCCACTC	GGCCTCACCCCATTTGATGT
Sox2	GGTTACCTCTTCCTCCCACTCCAG	TCACATGTG CGACAG GGGCAG
Nanog	AGGGTCTGCTACTGAGATGCTCTG	CAACCACTGGTTTTTCTGCCACCG
Gata4	CACCCCAATCTCGATATGTTTGA	GCACAGGTAGTGTCCCGTC
Sox17	GATGCGGGATACGCCAGTG	CCACCTCGCCTTTCACCTTTA
Sox1	AGATGCACAACTCGGAGATCAG	GAGTACTTGTCCTTCTTGAGCAGC
neuroD	AACCGCATGCACGGGCTGAA	GTTGGCAGATGCGGGGGCAT
Mixl1	GTCTTCCGACAGACCATGTACC	CCCGCCTTGAGGATAAGGG
Brachyury	GCTTCAAGGAGCTAACTAACGAG	CGTCACGAAGTCCAGCAAGA

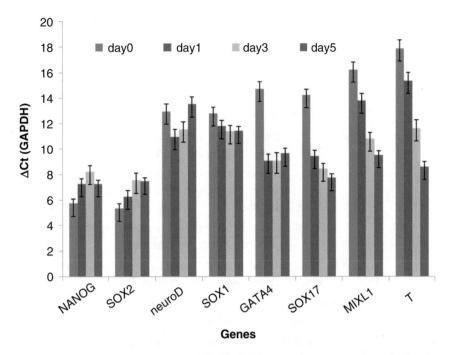

Fig. 4 The mRNA levels of pluripotent genes (NANOG, SOX2), ectoderm genes (neuroD, Sox1), endoderm genes (Gata4, Sox17), and mesoderm genes (Mixl1, T) in RCCS formed EBs were assessed by real-time PCR at day 0, 1, 3, and 5. mRNA expression of each gene relative to that of GAPDH was calculated using the ΔCt method. Mean and SEM from three separate experiments are shown

3.6 Analysis of EBs Formed in RCCS: Preparation of Slides and Immunostaining

1. After 6 days of culture, the EBs in RCCS are collected and prepared for immunostaining.

2. Transfer EBs from RCCS to 1.5 ml microcentrifuge tube and place the tube at static condition for 10 min to allow the EBs to sink to the bottom.

3. Remove medium from the tube and wash the EBs with PBS once.

4. Then fix with 4 % PFA at room temperature for 30 min and wash the EBs once with PBS.

5. Prepare 1.5 % aqueous agar solution and place the fixed and washed EBs into the tube containing the agar solution (1.5 % aqueous agar solution). The EBs are then centrifuged at $200 \times g$ for 5 min. After discarding the supernatant, 1.5 % aqueous agar solution is added and let harden in a refrigerator (4 °C) for 30 min.

6. After 30 min, the EBs are enclosed as a pellet in hardened agar blocks.

7. The agar blocks are embedded and frozen in OCT (Tissue Tek). Frozen tissue sections are cut at 10 μm on Leica cryostat and mounted on SuperFrost Plus slides (Fisher) as previously described [18].

8. Let the slides dry for 10–20 min at room temperature and fix the slides with 4 % PFA at room temperature for 15 min.

9. Wash the slides twice with PBS for 5 min and permeabilize for 15 min with 0.2 % Triton X-100 at room temperature.

10. Block the slides for 1 h at 37 °C with blocking solution (*see* **Note 10**).

11. Incubate the slides with primary antibody at 4 °C (or few hours at RT with less antibody) in blocking solution.

12. Wash the slides twice for at least 15 min with 0.2 % Triton X-100 in order to reduce the background and wash once for 15 min in PBS.

13. Incubate the slides with secondary antibody in blocking solution for 1 h at room temperature.

14. Wash the slides twice for at least 15 min with 0.2 % Triton X-100 and wash once for 5 min in PBS.

15. Add nuclear stain solution to the slides and incubate for 5 min at room temperature.

16. Wash the slides twice for 5 min in PBS.

17. Put one drop of mounting medium, add coverslip, and seal with nail polish.

18. Observe the slides and acquire images under Nikon Ti microscope (Fig. 5).

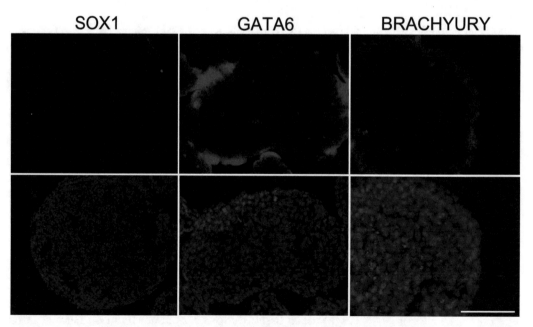

Fig. 5 Immunocytochemical staining of ectoderm marker (SOX1), endoderm marker (GATA6), and mesoderm marker (brachyury) in RCCS formed EBs at culture of day 4. *Blue*, nuclei. Bars: 100 μm

4   Notes

1. It is very important to use high-quality FBS and keep the same FBS lot throughout the cell culture. The quality of FBS should be tested before being used in large scale to culture mESCs. Different lots have differential effects on cell growth rate which in turn affects the EB formation and differentiation.

2. All disposable vessels are for single use; avoid reusing or resterilizing them.

3. Solubilization of paraformaldehyde powder needs low heat (52–56 °C), so do not heat the solution above 56 °C for a prolonged period of time, and adjust pH of 4 % paraformaldehyde solution (pH 7.2–7.4) with 1 N sodium hydroxide solution.

4. Importantly, to purify more mESCs separated from inactivated MEF, it is very important to control the attachment time of cell suspensions (no less than 30 min and no more than 1 h), and the cells should be observed every 5 min under microscope after 30 min attachment.

5. Check the vessel and be certain that valves and fill port are closed properly, ensuring that culture medium is not leaking from the vessel.

6. To get the higher EB concentration and homogenous EBs, we suggest the rotation speed of bioreactor in the range of 15–20 rpm.

7. After completion of medium change, make sure bubbles are removed.

8. In order to preserve RNA samples, which are very vulnerable to degradation at room temperature, do not wash cells before addition of TRIzol® Reagent to avoid increased chance of mRNA degradation.

9. Homogenize RNA samples using power homogenizer at room temperature to permit complete dissociation of the nucleoprotein complex.

10. Attention: do not let the frozen sections dry out in 37 °C and it should be carried out in a humidified chamber all the time to avoid drying of the sample after the slides have been fixed by 4 % PFA.

Acknowledgments

This work has been funded by National Basic Research Program of China (2011CB710905), Strategic Priority Research Program of the Chinese Academy of Sciences (XDA04020419, XDA04020202-20 and XDA01010202), the Chinese Manned Space Flight Technology Project (TZ-1) and the NSFC Grant (31471287).

References

1. Wu SM, Hochedlinger K (2011) Harnessing the potential of induced pluripotent stem cells for regenerative medicine. Nat Cell Biol 13:497–505
2. Hochedlinger K, Jaenisch R (2003) Nuclear transplantation, embryonic stem cells, and the potential for cell therapy. N Engl J Med 349:275–286
3. Dang SM, Kyba M, Perlingeiro R et al (2002) Efficiency of embryoid body formation and hematopoietic development from embryonic stem cells in different culture systems. Biotechnol Bioeng 78:442–453
4. Carpenedo RL, Sargent CY, McDevitt TC (2007) Rotary suspension culture enhances the efficiency, yield, and homogeneity of embryoid body differentiation. Stem Cells 25:2224–2234
5. Chen X, Xu H, Wan C et al (2006) Bioreactor expansion of human adult bone marrow-derived mesenchymal stem cells. Stem Cells 24:2052–2059
6. King JA, Miller WM (2007) Bioreactor development for stem cell expansion and controlled differentiation. Curr Opin Chem Biol 11:394–398
7. Rungarunlert S, Techakumphu M, Pirity MK et al (2009) Embryoid body formation from embryonic and induced pluripotent stem cells: benefits of bioreactors. World J Stem Cells 1:11–21
8. Kim JB, Stein R, O'Hare MJ (2004) Three-dimensional in vitro tissue culture models of breast cancer—a review. Breast Cancer Res Treat 85:281–291
9. Mitteregger R, Vogt G, Rossmanith E et al (1999) Rotary cell culture system (RCCS): a new method for cultivating hepatocytes on microcarriers. Int J Artif Organs 22:816–822
10. Villanueva I, Klement BJ, von Deutsch D et al (2009) Cross-linking density alters early metabolic activities in chondrocytes encapsulated in poly(ethylene glycol) hydrogels and cultured in the rotating wall vessel. Biotechnol Bioeng 102:1242–1250
11. Li S, Ma Z, Niu Z et al (2009) NASA-approved rotary bioreactor enhances proliferation and osteogenesis of human periodontal ligament stem cells. Stem Cells Dev 18:1273–1282
12. Lei X, Ning L, Cao Y et al (2011) NASA-approved rotary bioreactor enhances proliferation of human epidermal stem cells and supports formation of 3D epidermis-like structure. PLoS One 6:e26603
13. Lei X, Deng Z, Zhang H et al (2014) Rotary suspension culture enhances mesendoderm differentiation of embryonic stem cells through modulation of Wnt/beta-catenin pathway. Stem Cell Rev 10:526–538
14. Walker E, Ohishi M, Davey RE et al (2007) Prediction and testing of novel transcriptional networks regulating embryonic stem. Cell Stem Cell 1:71–86
15. Liu S, Wang X, Zhao Q et al (2015) Senescence of human skin-derived precursors regulated by Akt-FOXO3-p27(KIP(1))/p15(INK(4)b) signaling. Cell Mol Life Sci 72:2949–2960
16. Larionov A, Krause A, Miller W (2005) A standard curve based method for relative real time PCR data processing. BMC Bioinformatics 6:62
17. Livak KJ, Schmittgen TD (2001) Analysis of relative gene expression data using real-time quantitative PCR and the 2(-Delta Delta C (T)) method. Methods 25:402–408
18. Deng Z, Lei X, Zhang X et al (2015) mTOR signaling promotes stem cell activation via counterbalancing BMP-mediated suppression during hair regeneration. J Mol Cell Biol 7:62–72

Methods in Molecular Biology (2016) 1502: 77–86
DOI 10.1007/7651_2016_338
© Springer Science+Business Media New York 2016
Published online: 01 April 2016

Expansion of Human Mesenchymal Stem Cells in a Microcarrier Bioreactor

Ang-Chen Tsai and Teng Ma

Abstract

Human mesenchymal stem cells (hMSCs) are considered as a primary candidate in cell therapy owing to their self-renewability, high differentiation capabilities, and secretions of trophic factors. In clinical application, a large quantity of therapeutically competent hMSCs is required that cannot be produced in conventional petri dish culture. Bioreactors are scalable and have the capacity to meet the production demand. Microcarrier suspension culture in stirred-tank bioreactors is the most widely used method to expand anchorage dependent cells in a large scale. Stirred-tank bioreactors have the potential to scale up and microcarriers provide the high surface–volume ratio. As a result, a spinner flask bioreactor with microcarriers has been commonly used in large scale expansion of adherent cells. This chapter describes a detailed culture protocol for hMSC expansion in a 125 mL spinner flask using microcarriers, Cytodex I, and a procedure for cell seeding, expansion, metabolic sampling, and quantification and visualization using microculture tetrazolium (MTT) reagent.

Keywords: Mesenchymal stem cell, MSC, Bioreactor, Microcarrier, Spinner flask, Cell culture, Cell expansion

1 Introduction

Human mesenchymal stem cells (hMSCs) isolated from bone marrow and other tissues sources are an important candidate in cell therapy because they possess several important properties including self-renewability, high differentiation capabilities with specialized functions, and secretions of extracellular matrix, growth factors, cytokines, and chemokines [1–3]. The success of hMSC in clinical application and regenerative medicines require a large cell number, and the production of therapeutically competent hMSCs is important in hMSC clinical translation [4, 5]. Unlike human hematopoietic stem cells, hMSCs are anchorage-dependent cells and demand culture devices to provide the surface for attachment. Although multilayer vessels, also called Cell Factory™, can produce ten times of cells than regular petri dish culture due to the larger surface provided, further increase in production requires bioreactors that are capable of much higher batch capacity and space efficiency. Microcarriers are developed for anchorage-dependent cells to attach and grow and have been widely used in stirred-tank bioreactors in large scale cell expansion. Microcarriers with 100–300 μm in diameter provide the high surface-to-volume ratio which can culture more cells per unit volume [6, 7]. Many types of commercial microcarriers with a variety of properties have been designed and optimized for hMSC expansion with around 3 × 10^5 cells/mL at harvest cell density [4], thus reaching at least 10^{11} cells for harvest in a 1000 L bioreactor [8, 9]. In addition, the marine impeller agitation can maintain the homogeneous microcarrier distribution and accelerate oxygen and nutrition transport.

In the laboratory scale, hMSC expansion on microcarriers in stirred-tank bioreactors of 3–5 L bioreactors with 2–2.5 L working volume have been reported [6, 10–14]. While studies have shown that microcarrier suspension culture are exposed to the high shear stress due to the agitation in stirred-tank bioreactors [15, 16], and that the fluid flow in the bioreactor may influence hMSC fate [8, 17–20], microcarrier culture in stirred-tank bioreactors is still the most commonly used method to expand hMSCs in bioreactors. Microcarrier bioreactor is simple to be installed, autoclaved, controlled, sampled, and cleaned. The suspension of microcarriers is easily controlled by the agitation speed through the stirrer platform. Cell sampling can be done daily, which means the cell growth can be directly monitored during the culture time. More than above, the affordable cost of equipment as well as the potential design that can be easily applied to larger scales all explain why the microcarrier suspension in stirred-tank bioreactors culture appeals to the people who are interested in hMSC expansion in a large scale.

Spinner flask bioreactor (SFB) is the most common bioreactor used in laboratory scale for microcarrier culture. The SFB is typically a cylindrical glass vessel with two side arms and houses a shaft, a blade, and a magnetic stir bar, although some commercial disposable SFBs are made of plastic. SFB is small and can be set on a stirrer platform in a standard 37 °C, 5 % CO_2 incubator, without being necessarily connected with electronic sensors or probes, but the electronic probes are still able to be installed through the ports at the side arms if necessary. Direct cell counting, nucleus content, and metabolic activity are all able to quantify the cell population. Microculture tetrazolium (MTT) assay is developed as a quantitative colorimetric assay of metabolic activity for mammalian cell survival and proliferation, and has been widely used to test the hMSC growth in microcarrier culture [7, 21–23]. This protocol describes in detail how to expand hMSCs in a 125 mL SFB (Fig. 1a up) with commercial microcarriers, Cytodex I (Fig. 1a down), and measure the cell growth with MTT assay.

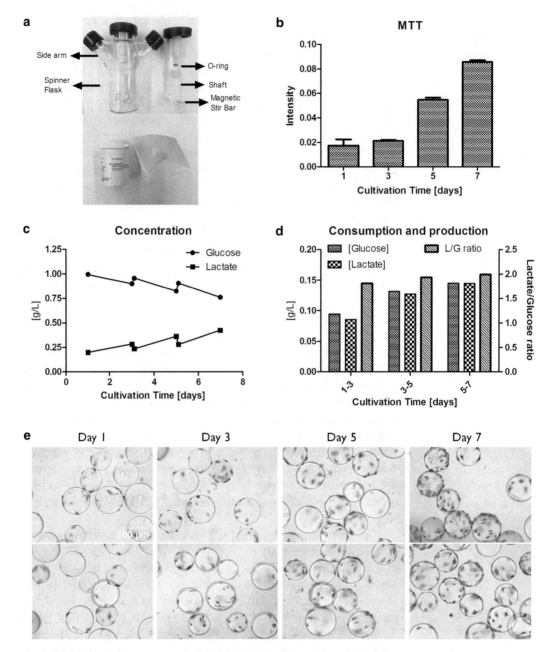

Fig. 1 (**a**) Spinner flask bioreactor and internal components (*top*). Cytodex I microcarriers (*bottom*). (**b**) hMSC proliferation measured by MTT intensities at day 1, 3, 5, and 7. (**c**) Concentration of glucose and lactate in the media over 7 day culture period. (**d**) The glucose consumption and lactate production with the lactate/glucose ratio over 7 day culture period. (**e**) MTT staining of hMSCs on Cytodex I at day 1, 3, 5, and 7

2 Materials

2.1 Preparing the Media

1. The complete cell culture media for hMSC expansion in petri dish or SFB culture is composed of 10.08 g Minimum Essential Medium Alpha Medium (Life Technologies, #12000-063) with 2.2 g sodium bicarbonate, 10 mL penicillin–streptomycin (Life Technologies, #15070-063), and 100 mL fetal bovine serum in 1 L.

2. Use a sterile filter with 0.2 μm pore size to filter the fresh media into a sterile bottle.

3. Store the bottle in a 4 °C refrigerator. Pre-warm it to 37 °C before use.

2.2 Preparing MSCs for Seeding in SFB

1. A standardized frozen vial of 10^6 hMSCs derived from human bone marrow is obtained from the Tulane Center for Gene Therapy and is used in this protocol.

2. Immerse the frozen vial in 70 % ethanol until melting. Open the vial in a biological safety cabinet and aspirate the solution into a sterile 15 mL centrifuge tube with 10 mL media. Centrifuge at $500 \times g$ for 5 min and confirm the cell pellet is visible. Aspirate the supernatant solution and replace with 5 mL fresh media. Resuspend the cells before transferring to 150 mm petri dish with 1 mL hMSCs and 25 mL fresh media per dish.

3. Culture them in a standard humidified 37 °C, 5 % CO_2 incubator and change media every 3 days.

4. Harvest them when cells reach 80 % confluence under microscopic observation.

5. Aspirate the media and use 10 mL phosphate-buffered saline (PBS) per dish to wash away the media. Use 4 mL of 0.25 % trypsin and 1 mM ethylenediaminetetraacetic acid (EDTA) solution per dish to rinse cells and then incubate for 7 min to detach cells. Confirm cells are rounded and floating under a microscope. Wash the suspended cells with fresh media. Centrifuge the cells down at $500 \times g$ for 5 min and replace with fresh media.

6. Resuspend and well mix the cells and take out 100 μL with 100 μL trypan blue (Life Technologies, # 15250-061) to count the cell number using a hemocytometer (*see* **Note 1**).

7. hMSCs are ready to be transferred into a SFB with the target cell number which can be obtained by calculating the cell density from the counting.

2.3 Preparing the Spinner Flask

1. A 125 mL SFB (Wheaton, #356876) is used to expand MSCs in this protocol. The SFB internal surface is rinsed with 5 mL Sigmacote (Sigma, #SL2). Remove excess solution and leave the siliconized SFB to air-dry in a hood overnight (*see* **Note 2**).

2. Wash the SFB internal surface with deionized water for three times before use.

3. Correctly install the impeller and adjust the location of the paddles to an appropriate height. The magnetic stir bar should be exactly in the middle (*see* **Note 3**).

4. Test the agitation of the impeller with 100 mL deionized water in the SFB on a stirrer platform. If the agitation is not fluent and stable, repeat the steps in Section 2.3 #3 and #4.

2.4 Preparing the Microcarriers

Cytodex 1 microcarriers (GE Healthcare Life Sciences, #17-0448-01) are used in this protocol as an example to expand MSCs in a microcarrier suspension system. 0.5 g dry microcarriers are immersed and hydrated in Ca^{2+} and Mg^{2+} free PBS for 4 h under room temperature. Gently decant the supernatant and replace with fresh Ca^{2+} and Mg^{2+} free PBS for twice (*see* **Note 4**). These microcarriers are ready to use.

2.5 Preparing the Sterile Culture System

Transfer the hydrated microcarriers with 30 mL PBS in the siliconized SFB. Close the cap but leave the cap loose. Cover with aluminum foil and label with autoclave tape. Autoclave the whole device (*see* **Note 5**).

2.6 Preparing MTT Assay

1. Thiazolyl blue tetrazolium bromide (Sigma-Aldrich, #M5655) is used to make MTT working solution with the concentration of 5 mg/mL (*see* **Note 6**).

2. A 96-well plate with transparent flat bottom is used for measuring MTT intensity in a colorimetric plate reader.

3. Dimethyl sulfoxide (DMSO) solution is prepared for dissolving the MTT dye from cells for colorimetric measurement.

3 Methods

All the procedures are described in the methods step by step. 100 mL working volume is operated in a 125 mL SFB for the total cultivation time of 7 days.

3.1 Seeding BMSCs into the SFB

1. After autoclave, carefully move the SFB into a biological safety cabinet for further operation. Gently aspirate the supernatant PBS out without aspirating microcarriers. Add fresh media to wash the microcarriers for 3 min and wait for 5 min to let the microcarriers settle down and then aspirate the supernatant media.

2. Add 40 mL fresh media in the SFB and put the SFB on a magnetic stirrer (Micro-Stir, Wheaton, #W900700-A) in a 37 °C, 5 % CO_2 incubator. Set agitation speed to 30 rpm to suspend the microcarriers for 2 h. Make sure all the surface of microcarriers is exposed to the media.

3. Harvest MSCs from plate culture and the cell number is determined by counting with a hemocytometer. Transfer 2×10^6 cells with 10 mL media into the SFB to reach a total volume of 50 mL (*see* **Note 7**).

4. Put the SFB on the stirrer in a 37 °C, 5 % CO_2 incubator. The intermittent agitation is recommended for the higher cell attachment efficiency and even cell distribution in the initial stage and can be set from the program of the magnetic stirrer. A cycle with 5 min ON at 30 rpm and 30 min OFF is used for a total of 20 cycles. After the intermittent agitation, continuous agitation is at 30 rpm in the period of cultivation time. All the microcarriers should be well suspended without sedimentation. If not, the agitation speed should be increased (*see* **Note 8**).

5. The culture volume is increased to 100 mL after 1 day of seeding. The less volume in the beginning stage is to increase the possibility of cell–microcarrier collision and thus enhance cell attachment.

3.2 Bioreactor Operation and Sample Collection

3.2.1 Day 1

1. First, the culture volume is increased to 105 mL.

2. Move the stirrer platform and the SFB in the biological safety cabinet. Keep the continuous agitation speed at 30 rpm, and collect 1 mL per time through the side port of the SFB into a 1.5 mL microcentrifuge tube for four times totally. Three samples are prepared for MTT measurement to monitor cell growth and the other is prepared for MTT staining.

3. Turn off the agitation and wait for 5 min to let the microcarriers settle down. Take out 1 mL media from the supernatant for glucose and lactate analysis.

4. Move the stirrer platform and the SFB back to the 37 °C, 5 % CO_2 incubator.

3.2.2 Day 3

1. Repeat the steps in Section 3.2.1 #2 and #3.

2. After sampling, change the media. Turn off the agitation and wait for 5 min to let the microcarriers settle down. Remove 45 mL media from the supernatant. Refill 51 mL fresh media into the SFB and turn on the agitation to mix the media for 10 min.

3. Repeat the steps in Section 3.2.1 #3 and #4.

3.2.3 Day 5

1. Repeat the steps in Section 3.2.1 #2 and #3.

2. Repeat the steps in Section 3.2.2 #2.

3. Repeat the steps in Section 3.2.1 #3 and #4.

3.2.4 Day 7

1. Repeat the steps in Section 3.2.1 #2 and #3.

3.3 MTT Measurement

1. Gently pipette one of the collected samples of 1 mL to suspend the microcarriers well and take out 900 μL into a new 1.5 mL microcentrifuge tube with 100 μL MTT working solution to reach the concentration of 10 % MTT working solution. Slightly pipette twice to mix the MTT solution.

2. Repeat the same procedures for the other two samples. Then incubate the three microcentrifuge tubes in a 37 °C, 5 % CO_2 incubator for 45 min.

3. After incubation, the microcarriers should show some purple color to naked eye. Centrifuge the three samples at $500 \times g$ for 5 min, and remove the supernatant and replace with fresh PBS to resuspend the microcarriers. Centrifuge again and remove the PBS and then replace with 350 μL DMSO solution to extract the MTT dye out of cells. Slightly pipette twice to mix DMSO solution. Wait for 3 min to see if all the microcarriers become colorless. Otherwise, gently pipette and wait longer.

4. Centrifuge at $500 \times g$ for 5 min to settle down the microcarriers. Take 100 μL from the supernatant into a transparent flat-bottom 96-well plate in triplicate for each sample, and a triplicate of 100 μL fresh DMSO as background.

5. MTT intensity is measured by a microplate reader at absorbance wavelength of 490 nm with background subtraction and the results are shown on Fig. 1b.

3.4 MTT Staining

1. Follow the same steps in Section 3.3 #1. Incubate the sample in a 37 °C, 5 % CO_2 incubator for 45 min.

2. Gently resuspend the microcarriers and take 100 μL solution with microcarriers in a transparent flat-bottom 96-well plate in triplicate.

3. Randomly capture images by an Olympus IX70 microscope at 10× as Fig. 1e (*see* **Note 9**).

3.5 Glucose and Lactate Analysis

1. After sampling 1 mL media from the SFB, centrifuge at $500 \times g$ for 5 min. Carefully take out 500 μL supernatant media for measuring glucose and lactate level by a YSI-2700 analyzer as Fig. 1c (*see* **Note 10**). Subtraction of the concentration in a period of cultivation time represents the consumption of glucose and the production of lactate (Fig. 1d left *y*-axis).

2. The lactate versus glucose ratio can be obtained by doubling the lactate production over the glucose consumption (Fig. 1d right *y*-axis).

4   Notes

1. If the cell number counted from the four areas of a hemocytometer is less than 100, it means the cell density is too low to count a more accurate number. In order to get the higher cell density solution for counting, centrifuge again to get rid of some media, and then resuspend the cells and take samples for counting, until the counting number is higher than 100 cells.

2. Sigmacote treatment is used to perform the hydrophobic surface to avoid cell attachment on the glass surface during the culture. Therefore, cells can all attach on the microcarriers, and microcarriers with cells will not stick on the internal surface of the SFB.

3. The height of the paddle depends on the length of the shaft, and the length of the shaft is determined by the location of the O-ring which can be seen after screwing out the impeller part from the shaft. The end point of the shaft should not touch the bottom wall of the SFB. Make sure the magnetic stir bar is stuck to the center and cannot slip off during the agitation.

4. When wash Cytodex I microcarriers with PBS, mix or slight vortex the microcarriers every time for complete suspension. There may be few microcarriers cannot immerse and submerge in PBS. Just wash them out. Incubation of the microcarriers for 4 h is to confirm the surface of the microcarriers is thoroughly hydrated.

5. The reason for leaving the cap loose is to let steam penetrate into the SFB during autoclave. Otherwise, the glass vessel may break under high pressure.

6. The aliquots of MTT working solution can be stored in $-20\,^{\circ}\mathrm{C}$ from light because of its light sensitivity.

7. The reduced volume for the initial seeding stage is to increase the cell and microcarrier density and thereby enhance the possibility of cell-microcarrier collision and attachment.

8. The intermittent agitation is set for initial cell attachment. Cells gradually settle down and attach to the top surface of the microcarriers at the agitation cycle time OFF, while resuspend the cells and rotate the microcarriers for even cell attachment at the agitation cycle time ON. The duration for the cycle time ON and OFF can be adjusted. If the uneven cell distribution is observed after seeding, reduce OFF time and increase ON time in the next trial. If the percentage of attached cells is low and many cells are still suspended after seeding stage, increase OFF time and reduce ON time in the next trial.

9. When collecting images for the MTT stained microcarriers, do not put too many microcarriers into 96-well plate. The crowded microcarriers cannot be laid on the same surface, thus generating the problem of focus.

10. The reason for centrifuging the microcarrier down after sampling the media from the SFB is to avoid microcarriers getting into the YSI analyzer, which may result in inaccurate reading and also damage the machine.

References

1. Daley GQ, Scadden DT (2008) Prospects for stem cell-based therapy. Cell 132:544–548

2. Koga H, Engebretsen L, Brinchmann JE, Muneta T, Sekiya I (2009) Mesenchymal stem cell-based therapy for cartilage repair: a review. Knee Surg Sports Traumatol Arthrosc 17:1289–1297

3. Meijer GJ, de Bruijn JD, Koole R, van Blitterswijk CA (2007) Cell-based bone tissue engineering. PLoS Med 4:e9

4. Chen AK-L, Reuveny S, Oh SKW (2013) Application of human mesenchymal and pluripotent stem cell microcarrier cultures in cellular therapy: achievements and future direction. Biotechnol Adv 31:1032–1046

5. Jung S, Panchalingam KM, Wuerth RD, Rosenberg L, Behie LA (2012) Large-scale production of human mesenchymal stem cells for clinical applications. Biotechnol Appl Biochem 59:106–120

6. Merten O-W (2015) Advances in cell culture: anchorage dependence. Philos Trans R Soc Lond B Biol Sci 370:20140040

7. Malda J, Frondoza CG (2006) Microcarriers in the engineering of cartilage and bone. Trends Biotechnol 24:299–304

8. Ma T, Tsai A-C, Liu Y (2015) Biomanufacturing of human mesenchymal stem cells in cell therapy: influence of microenvironment on scalable expansion in bioreactors. Biochem Eng J. doi:10.1016/j.bej.2015.07.014

9. Rowley J, Abraham E, Campbell A, Brandwein H, Oh S (2012) Meeting lot-size challenges of manufacturing adherent cells for therapy. BioProcess Int 10:7

10. Kehoe D, Schnitzler A, Simler J, DiLeo A, Ball A (2012) Scale-up of human mesenchymal stem cells on microcarriers in suspension in a single-use bioreactor. BioPharm Int 25:28–38

11. Jing D, Sunil N, Punreddy S, Aysola M, Kehoe D, Murrel J, Rook M, Niss K (2013) Growth kinetics of human mesenchymal stem cells in a 3-L single-use, stirred-tank bioreactor. BioPharm Int 26:28–38

12. Rafiq QA, Brosnan KM, Coopman K, Nienow AW, Hewitt CJ (2013) Culture of human mesenchymal stem cells on microcarriers in a 5 l stirred-tank bioreactor. Biotechnol Lett 35:1233–1245

13. van Eikenhorst G, Thomassen YE, van der Pol LA, Bakker WA (2014) Assessment of mass transfer and mixing in rigid lab-scale disposable bioreactors at low power input levels. Biotechnol Prog 30:1269–1276

14. Tsai A-C, Liu Y, Ma T (2015) Expansion of human mesenchymal stem cells in fibrous bed bioreactor. Biochem Eng J. doi:10.1016/j.bej.2015.09.002

15. Ismadi M-Z, Hourigan K, Fouras A (2014) Experimental characterisation of fluid mechanics in a spinner flask bioreactor. Processes 2:753–772

16. Liovic P, Šutalo ID, Stewart R, Glattauer V, Meagher L (2012) Fluid flow and stresses on microcarriers in spinner flask bioreactors. In: Proceedings of the 9th international conference on CFD in the minerals and process industries

17. Scaglione S, Wendt D, Miggino S, Papadimitropoulos A, Fato M, Quarto R, Martin I (2008) Effects of fluid flow and calcium phosphate coating on human bone marrow stromal cells cultured in a defined 2D model system. J Biomed Mater Res A 86:411–419

18. Leung HW, Chen A, Choo AB, Reuveny S, Oh SK (2010) Agitation can induce differentiation of human pluripotent stem cells in microcarrier cultures. Tissue Eng Part C Methods 17:165–172

19. Yi W, Sun Y, Wei X, Gu C, Dong X, Kang X, Guo S, Dou K (2010) Proteomic profiling of human bone marrow mesenchymal stem cells under shear stress. Mol Cell Biochem 341:9–16

20. Zhao F, Chella R, Ma T (2007) Effects of shear stress on 3-D human mesenchymal stem cell construct development in a perfusion bioreactor system: experiments and hydrodynamic modeling. Biotechnol Bioeng 96:584–595

21. Mosmann T (1983) Rapid colorimetric assay for cellular growth and survival: application to proliferation and cytotoxicity assays. J Immunol Methods 65:55–63

22. Sart S, Schneider Y-J, Agathos SN (2009) Ear mesenchymal stem cells: an efficient adult multipotent cell population fit for rapid and scalable expansion. J Biotechnol 139:291–299

23. Frauenschuh S, Reichmann E, Ibold Y, Goetz PM, Sittinger M, Ringe J (2007) A microcarrier-based cultivation system for expansion of primary mesenchymal stem cells. Biotechnol Prog 23:187–193

Methods in Molecular Biology (2016) 1502: 87–102
DOI 10.1007/7651_2015_314
© Springer Science+Business Media New York 2015
Published online: 19 February 2016

Large-Scale Expansion and Differentiation of Mesenchymal Stem Cells in Microcarrier-Based Stirred Bioreactors

Sébastien Sart and Spiros N. Agathos

Abstract

Mesenchymal stem cells (MSCs) have emerged as an important tool for tissue engineering, thanks to their differentiation potential and their broad trophic activities. However, for clinical purposes or for relevant in vitro applications, large quantities of MSCs are required, which could hardly be reached using conventional cultivation in plastic dishes. Microcarriers have high surface to volume ratio, which enables the easy scale-up of the expansion and differentiation of MSCs. In addition, the agitation in stirred tank bioreactors limits the diffusion gradient of nutrients or morphogens, thus providing a physiologically relevant environment to favor MSC production at large scale. This work describes a simple method for the mass expansion and differentiation of MSCs, including the procedures to monitor the proliferation, metabolic status and phenotype of MSCs during suspension culture. Moreover, this work proposes suitable materials for cGMP compliant culture conditions enabling the clinical grade production of MSCs.

Keywords: Mesenchymal stem cells, Microcarriers, Spinner flasks, cGMP

1 Introduction

Mesenchymal stem cells (MSCs) are primary candidates for tissue engineering purposes, thanks to their differentiation potential and their broad trophic activities [1–4]. MSCs are listed in more than 500 clinical trials, which include the treatment of osteogenesis imperfecta, diabetes, heart infarction, etc. (www.clinicaltrials.gov). MSCs have been recently characterized in vivo as pericytes [5, 6], a cell population which plays an important role in maintaining various tissue functions such as vasomotion, blood vessel maturation, and regulation of extracellular matrix turnover, through mechanisms like paracrine signaling among others [7]. The re-creation of their original niche during in vitro culture is essential to maintain the MSC secretory competence. On the other hand, in vitro engineering the microenvironment of MSCs can promote their differentiation towards specialized cell types, which can have broad applications for drug screening and disease modeling in vitro [8, 9] or for tissue engineering in vivo.

The use of MSCs in therapy or for in vitro applications requires large amounts of cells, which could hardly be reached using conventional 2D static culture. For instance, 20–200 million cells are required for treatment of heart diseases, while 3000–6000 million MSCs have been transplanted for curing osteogenesis imperfecta [10]. In addition, to ensure the safety and the efficacy of MSC transplantation, the scalable production of MSCs in a cGMP compliant manner is critically required [11]. Moreover, the efficient priming of MSCs for the enhancement of their secretory activities or for improving the extent of differentiation requires the controlled application of biochemical and biomechanical cues, which could be achieved using well-designed bioreactors.

Microcarriers (MCs) were first used in 1967 for the culture of anchorage dependent mammalian diploid cells [12]. Since then, different adherent cell types (e.g., Vero, MDCK) have been cultivated in MC-based stirred tank bioreactors for the clinical scale production of biologicals, including vaccines, viral vectors, or recombinant proteins under cGMP conditions. The culture of MSCs on MCs has been demonstrated in the last decade, and the successful production of up to 7×10^8 functional MSCs in this system has been reported [10, 13, 14]. The agitation of the culture medium in spinner flasks (SFs) improves the mass transfer and limits the gradient diffusion of morphogens and nutrients (e.g., glucose and oxygen). Thus, these culture conditions ensure the homogeneity of the MSC biochemical environment. In addition, MCs can serve as biomaterial scaffolds for cell delivery in-situ [15]. Together, the MC-based stirred bioreactor platform emerges as the easiest way to achieve large-scale expansion and the differentiation of MSCs for clinical applications.

This work describes a method to successfully expand and recover large numbers of MSCs in a functional status suitable for in vivo transplantation or in vitro applications. In addition, this work describes the procedures for monitoring the proliferation, the metabolic status, and the phenotype of MSCs during suspension culture. Finally, this work proposes suitable materials for cGMP compliant culture conditions enabling the clinical grade production of MSCs at large scale.

2   Materials

1. *Spinner flasks*: SFs can be obtained as "ready to use disposable spinner flasks" from Corning Life Science: 500 mL (Corning ref. #3578), 1 L (Corning ref. #3580), 3 L (Corning ref. #3581); or as "glass reusable spinner flasks" from Bellco Glass: Bell-Flo Complete, 500 mL–70 mm (ref. #:

1965-80575) 1 L (ref. #: 1965-81005) or 3 L (ref. #: 1965-83005).

2. Sigmacote® (SL2) can be obtained from Sigma-Aldrich.

3. Magnetic agitators can be obtained from Bellco Glass: Bell-stir Magnetic (ref. #: 7760-06005).

4. *Microcarriers*: MCs can be obtained from SoloHill® (BB-MF 3092 collagen coated, BB-MF 3093 Glass coated, BB-MF 3094 Plastic and Plastic Plus (copolymer), BB-MF 6889 FACT III, BB-MF 6915 ProNectin® F, BB-MF 10059 Hillex II); Percell-Biolytica (Cultispher®); GE-Healthcare Life Science (Cytopore-1, ref. #17-0911-01; Cytopore-2, ref. #17127101; Cytodex-1 ref. #17-0448-01; Cytodex-3, ref. #17-0485-01). Table 1 indicates the properties of the different types of commercially available MCs, and Fig. 1a shows different types of MCs (all reproduced from Sart et al. [13]) (*see* **Note 1**).

Table 1
Characteristics of various commercial MCs (reproduced from Sart et al. [13])

MC type	Charge	Diameter (µm)	Surface area (cm²/g)	Pore size	Coating
Biosilon (Nunc™) (Polystyrene)	No charge	160–300	255	Solid	No additional coating
Collagen (SoloHill) (polystyrene)	No charge	90–150	480	Solid	Type 1 porcine collagen (gelatin)
CultiSpher-S (Percell-Biolytica) (gelatin)	No charge	130–380	7500	20 µm	None
Cytopore 2 (GE Healthcare Life Sciences) (cellulose linked with DEAE)	Cationic (1.8 meq/g)	200–280	11,000	30 µm	No additional coating
Cytodex-3 (GE Healthcare Life Sciences) (dextran based)	No charge	141–211	2700	Solid	Cross-linked with gelatin
Cytodex-1 (GE Healthcare Life Sciences) (dextran linked with DEAE)	Cationic (1.4–1.6 meq/g)	190 ± 58	4400	Solid	No additional coating
DE-53 (Whatman) (cellulose linked with DEAE)	Cationic (1.8–2.2 meq/g)	35–40	6800	Solid	No additional coating

(continued)

Table 1
(continued)

MC type	Charge	Diameter (μm)	Surface area (cm²/g)	Pore size	Coating
DE-52 (Whatman) (cellulose linked with DEAE)	Cationic (0.88–1.08 meq/g)	35–40	6800	Solid	No additional coating
FACT III (SoloHill) (polystyrene)	Cationic	90–150	480	Solid	Type 1 porcine collagen (gelatin)
Fibra-Cel® (New Brunswick-Eppendorf) (PET polyester)	N.A.	6000	120	Porous	No additional coating
HILLEX® II (SoloHill) (modified polystyrene)	Cationic	160–200	515	Solid	No additional coating
Glass (SoloHill) (Cross-linked polystyrene coated with high silica glass)	No charge	125–212	360	Solid	No additional coating
MicroHex (Nunc™) (polystyrene)	N.A.	125–212	360	Solid	No additional coating
Plastic (SoloHill) (Polystyrene)	N.A.	90–150	480	Solid	No coating
PlasticPlus (SoloHill) (polystyrene)	Cationic	90–150	480	Solid	No coating
ProNectin®F-COATED (SoloHill) (polystyrene)	Cationic	90–150	480	Solid	Recombinant RGD-containing protein
RapidCell (MP biomedical) (glass)	No charge	150–210	325	Solid	No additional coating
Synthemax® II MCs (Corning) (polystyrene conjugated with peptides)	N.A.	125–212	360	Solid	Synthemax II (vitronectin peptide conjugated with carboxylic acrylate)
Tosoh 65 PR (Tosoh Bioscience) (hydroxylated methacrylate)	N.A.	65	4200	Solid	Protamine sulfate (cationic peptide)
Tosoh 10 PR (Tosoh Bioscience) (hydroxylated methacrylate)	N.A.	10	90,000	Solid	Protamine sulfate (cationic peptide)

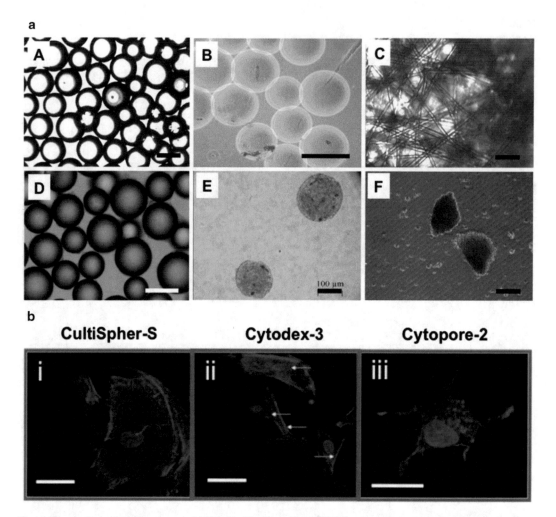

Fig. 1 Morphology of various types of MCs (**a**). (*A*) Glass (SoloHill) MCs as spherical beads; (*B*) Cytodex-3 as spherical beads; (*C*) Fibra-Cel® with porous microfibers; (*D*) Synthemax® II MCs with vitronectin peptide acrylate surface (adapted from Corning product insert, http://www.corning.com/lifesciences); (*E*) rat cartilage MSCs grown on CultiSpher-S; (*F*) PSC-derived extracellular matrix MCs. Scale bars = 100 μm. (**b**) Shape of cartilage-derived MSCs on (*i*) CultiSpher-S (high degree of spreading and well organized actin), (*ii*) Cytodex-3 (lower degree of spreading), and (*iii*) Cytopore (round shape and disorganized actin) (All reproduced from Sart et al. [13])

5. *Culture media for expansion and differentiation on MCs*: conventional expansion medium consists of α-MEM-containing low glucose (Life Technologies) supplemented with 10 % fetal bovine serum (FBS) (Life Technologies), which can serve as basal medium for osteogenic and adipogenic differentiation. For chondrogenic differentiation, the basal medium is DMEM containing low glucose and GlutaMAX that can be obtained from Life Technologies. The components of the various differentiation media (i.e., L-ascorbic acid and dexamethasone, glycerol phosphate, ITS premix, and TGF-β3, as

indicated in Section 3.2.2) can be obtained from Sigma-Aldrich. For cGMP compliant culture conditions, xeno-free expansion medium (e.g., StemPro® MSC SFM XenoFree), StemPro® kits for differentiation and TryplE™ for cell recovery can be obtained from Life Technologies.

6. *Reagents for the analysis of MSC adhesion, proliferation, and functions on MCs.* Propidium iodide, crystal violet, alcian blue, Oil red O, alizarin red can be obtained from Sigma-Aldrich. Primary and secondary antibodies, ELISA kits can be obtained from Abcam. Primers, Sybr® Green or TaqMan probes, and Phalloidin Alexa® 594 can be obtained from Life Technologies.

3   Methods

3.1   MSC Extraction, Subculture, and Minimal Characterization

MSCs can be isolated form various connective tissues such as umbilical cord, bone marrow, and adipose tissue:

3.1.1   MSC Isolation from Tissue Explants

1. The extraction of MSCs begins with tissue aspirate (e.g., bone marrow) or with explants (e.g., adipose tissue, Wharton jelly), which requires enzymatic digestion (e.g., using recombinant collagenase treatment).

2. Tissue aspirate or lysate is then plated at 2000–5000 cells/cm^2 in plastic dishes using DMEM supplemented with 10 % screened FBS (expansion medium), autologous serum, or serum/xeno free expansion medium.

3. Some fibroblastic colonies should appear within 2–3 days post-replating, which correspond to the MSC population.

3.1.2   MSCs from a Cell Bank

MSCs can be thawed from a cell bank usually placed in a nitrogen tank, where cells are stored at $-180\,^{\circ}$C in cryopreservation medium (usually containing 90 % FBS and 10 % DMSO; or a protein-free solution such as Cryostor® and HypoThermosol® FRS) [16]:

1. The cryovials are quickly thawed in a water bath set up at 37 °C.

2. The cryopreservation medium is equilibrated dropwise with expansion medium (with about 3–4 times the volume of the cryo-medium).

3. The cells are collected by centrifugation at 0.9 relative centrifugal force (RCF) for 5 min.

4. MSCs are replated at 2000–5000 cells/cm^2 on T-flasks in expansion medium.

3.1.3   MSC Subculture in Monolayer

1. Sub-confluence should be avoided to reduce premature senescence and to ensure that the cells retain their differentiation potential. For human MSCs, the senescence may occur after around 15–20 passages, significantly altering their therapeutic

functions [17, 18]. Thus, the culture of MSCs beyond 40 population doublings should be avoided.

2. The MSC population is regularly passaged when reaching 70 % confluence, using trypsin or TrypLE™ (Life Technologies).

3. The cells are re-seeded at 2000–5000 cells/cm^2 in expansion medium.

3.2 Characterization of MSC Population

To be defined as MSCs, the cell population must demonstrate (a) expression of specific membrane markers and (b) differentiation potential along osteogenic, adipogenic, and chondrogenic lineages [19].

3.2.1 Analysis of Membrane Marker Expression

1. MSCs are collected from the 2D culture vessels or from MCs as a single cell suspension.

2. MSCs are fixed in 4 % paraformaldehyde (PFA) solution.

3. MSCs are blocked in a 5 % FBS solution.

4. MSCs are incubated with anti-CD14, -CD19, -CD44, -CD90, -STRO-1, -CD146, -CD105, and -CD73 primary antibodies (usually diluted at 1:200 in staining buffer (SB) which contains 1 % FBS) for a minimum of 30 min at room temperature.

5. MSCs are washed in PBS and stained for 30 min with secondary antibodies (when non-conjugated primary antibodies are first used), which are diluted in SB at the same concentration as the primary antibodies.

6. After washing with PBS, MSCs are analyzed by flow cytometry (*see* **Note 2**).

3.2.2 MSC Differentiation

To demonstrate the differentiation potential along osteogenic, adipogenic, and chondrogenic lineages, MSCs from T-flasks or from MC cultures are seeded at specific concentrations in 2D or as 3D aggregates and are incubated with specific induction media for defined culture periods:

1. *For adipogenic differentiation*: MSCs are seeded at high concentration (about 1×10^4 cells/cm^2) in plastic dishes. The day after, seeding expansion medium is replaced by differentiation medium containing 3-isobutyl-1-methyl xanthine (0.5 mM), dexamethasone (1 M), and insulin (50 M). The differentiation medium is refreshed every 2 days and the cells are cultivated for a culture period of 21 days.

2. *For osteogenic differentiation*: MSCs are seeded at lower concentration (5×10^3 cells/cm^2) in plastic dishes. The day after, seeding expansion medium is replaced by differentiation medium containing 50 mM L-ascorbic acid plus 100 nM dexamethasone and 2 mM glycerol phosphate. The differentiation

medium is refreshed every 2 days, until day 21 of differentiation.

3. *For chondrogenic differentiation*: MSCs must be cultivated as aggregates. About 1×10^6 cells are pelleted by centrifugation, or, alternatively, MSC aggregates could be formed by hanging drops or in AggreWell™. The aggregates are then cultivated in serum free medium containing DMEM plus GlutaMAX, 1× ITS premix (insulin, transferrin, selenium), 50 mM L-ascorbic acid, 100 nM dexamethasone, and 10 ng/mL TGFβ-3. The cells are incubated in this medium for a total period of 14 days and it is exchanged with fresh medium every 4 days.

3.2.3 Quantification of MSC Differentiation

1. *Flow cytometry*: MSCs are harvested from the 2D culture vessels or from MCs. Different intracellular markers can be quantified for assessing MSC differentiation using conventional methods for immuno-staining (Section 3.2.1):

    (a) Osteogenic differentiation: RUNX-2, osteocalcin, osteo-pontin, BSP, etc.

    (b) Adipogenic differentiation: C/EBP-α, PPAR-γ, etc.

    (c) Chondrogenic differentiation: Collagen type II, Aggre-can, etc. (*see* **Note 3**)

2. *RT-qPCR*: The same markers as for flow cytometry can be used for the quantification of MSC differentiation at the transcriptional level. The total RNA can be directly recovered from MSCs on MCs using conventional RNA extraction reagents (e.g., TRIzol®, Superscript™ III CellsDirect cDNA synthesis System). RNA is reverse-transcribed and cDNA is amplified using dedicated primers, Sybr® Green or TaqMan® probes, following conventional RT-qPCR methods.

    The level of expression of differentiation markers must be compared to undifferentiated MSCs cultivated for similar time periods (*see* **Note 4**).

3. *Quantitative stainings* can be performed directly in the 2D culture vessel or on MCs (*see* **Note 5**):

    (a) Alizarin red enables the quantification of calcium deposition after osteogenic differentiation. After standard staining procedures [5], the dye can be extracted from the samples with a solution containing 20 % methanol plus 10 % acetic acid, for subsequent measurement with a spectrophotometer.

    (b) Alkaline phosphatase (ALP) is upregulated during the osteogenic differentiation. ALP can be detected after incubation with 5 mM *p*-nitrophenyl phosphate (a substrate for ALP) and can be quantified spectrophotometrically [20].

(c) Oil red O enables the staining by standard procedure of the intracellular lipid vacuoles [21], which appear during the adipogenic differentiation. After extraction with 100 % isopropanol, Oil red O can be spectrophotometrically measured.

(d) Alcian blue binds specifically to glycoaminoglycans (GAGs), which are upregulated during chondrogenic differentiation. After conventional staining [21], alcian blue can be spectrophotometrically measured following its extraction using a solution made of 6 M guanidine HCl.

3.2.4 Additional Characterization of the MSC Population

1. *CFU-F assay* (*see* **Note 6**): MSCs are collected from the culture vessels or MCs and are seeded at a suitable clonal density (about 2 cells/cm^2) in monolayer. After a 1-week culture period, MSCs are stained with 1 % crystal violet in 100 % methanol, and the number of colonies are counted. CFU-F efficiency is defined by:

   CFU-F efficiency = (counted colonies/seeded cell number) × 100

2. *Analysis of the MSC secretory functions*: After centrifugation, the MSC culture supernatant can be analyzed by ELISA for PGE2, nitrogen oxide (NO, an anti-inflammatory molecule), VEGF (a pro-angiogenic growth factor), FGF (a mitotic growth factor), NGF (a neurogenic growth factor), etc.

3.3 Spinner Flasks and Microcarrier Preparation: MSC Seeding in Spinner Flasks

3.3.1 Microcarrier (MC) Preparation

1. MCs (about 1–5 g/L) are hydrated by incubation in PBS overnight.

2. MCs are sterilized by autoclaving at 120 °C/1 bar for 20 min.

3. After cooling at room temperature, the MCs are coated with proteins from expansion medium containing FBS or with recombinant ECM proteins (recombinant fibronectin, collagen, CellStart™, etc.), if serum-free media are used.

3.3.2 Spinner Flasks (SFs) Preparation

- Reusable glass spinner flasks:

  1. SFs are siliconized overnight using Sigmacote®.

  2. SFs are washed at least five times with PBS to eliminate remaining traces of Sigmacote®.

  3. SFs are autoclaved at 120 °C/1 bar for 20 min.

  4. SFs are cooled down at room temperature prior to seeding with MSCs and MCs.

- Sterile single use spinner flasks can be used without any further treatment.

3.3.3 MSC Seeding on MCs in SF

1. MCs are seeded in SFs filled at one-half of the SF working volume.

2. The starting concentration of MSCs is about 5×10^4 cells/mL.

3. SF magnetic agitation is set up for intermittent stirring (3 min every 30 min) for 24 h at minimal speed (about 30 rpm).

4. SF magnetic agitation is set up for continuous low agitation (about 30 rpm) for the rest of the culture period.

3.4 Monitoring MSC Adhesion, Proliferation, and Phenotype in MC-Based Bioreactor

- *Quantification of the seeding efficiency (SE)*. After the 24 h cultivation under intermittent stirring in SFs:

  1. The MSCs attached on MCs are counted (*see* Section 3.4.2).

  2. The MSCs in suspension are counted (by trypan blue exclusion).

  3. The seeding efficiency is calculated as follows (*see* **Note 7**):

     SE = [(attached cells number)/(attached cell number + non-attached cell number)] × 100

3.4.1 Monitoring MSC Adhesion on MCs

- *The qualitative assessment of actin organization* can give a good indication how the cells adhere on the MCs:

  1. A sample of MSC-MC constructs is recovered from SFs.

  2. MSCs are fixed in 4 % PFA, then permeabilized in a 0.5 % Triton X-100 solution and blocked in 5 % FBS.

  3. Cells are stained for F-actin using a 1:50 Phalloidin Alexa® 594 solution diluted in 1 % FBS.

  4. After 30 min incubation, the cells are washed with PBS, counterstained with DAPI, and imaged using a confocal microscope (*see* **Note 8**).

3.4.2 Monitoring Cell Proliferation and Metabolic Status in SFs

Different methods can be used for direct cell counting on MCs, or indirectly through correlation with metabolic measurements. While the full digestion of MC by enzymes is the easiest method for cell collection and counting, crystal violet staining can be used for nuclei counting, when digestion of MC is hard to achieve.

1. *Crystal violet staining*: MCs are allowed to settle and the expansion medium is replaced by a solution of 0.1 % crystal violet in 1 M citric acid plus 0.1 % Triton X-100. After 1 h of incubation at 37 °C, the stained nuclei are counted using a hemacytometer [22].

2. *MC digestion and trypan blue counting*. For this application, the example is given of MSC counting on Cultispher-S. An aliquot containing suspended MCs from SFs is harvested. MCs are allowed to settle. After washing in PBS and total digestion with dispase grade II, trypsin 10×, or TypLE™, MSCs are

collected by centrifugation and counted by trypan blue exclusion.

3. *Cell cycle analysis*: MSCs are retrieved from MCs, then fixed and permeabilized with a 70 % ethanol solution at 4 °C for 40 min. The cells are then centrifuged and the pellet is resuspended in a solution containing 20 g/mL propidium iodide (Sigma-Aldrich) plus 0.5 mg/mL RNAse A (Sigma-Aldrich). The cells are incubated at 37 °C for 1 h. Finally, the cells are washed in PBS and analyzed by flow cytometry (*see* **Note 9**).

4. *Indirect monitoring of MSC expansion by measurement of the metabolic activity*

   The MTT assay measures the MSC metabolic activity. When compared to a standard curve from defined cell densities, the MTT assay enables to calculate MSC number without the need for cell recovery from MCs. For this assay, the expansion medium is exchanged to a 5 mg/mL of 3-(4,5-dimethylthiazol-2-yl)-2,5-diphenyltetrazolium bromide (MTT) solution. After 15–30 min incubation, the formazan crystals are hydrolyzed with dimethyl sulfoxide and the absorbance of the resulting violet solution is measured at 500 nm using a microplate reader [23].

   The measurement of glucose concentration in the culture supernatant using colorimetric kits, HPLC, or Raman spectrometry enables a similar quantification of cell proliferation, without the need for MSC separation from the MCs (*see* **Note 10**).

   Importantly, to ensure the efficient expansion of MSC at the undifferentiated state, the cells must be checked for their basic properties and phenotype, as described in Sections 3.2.1 and 3.2.3, following the specific indications for the MSC on MCs. For the quantification the differentiation potential after bioreactor expansion, MSCs could be extracted from MCs, replated in 2D cultures or directly differentiated on the MCs used for expansion (*see* Section 3.6).

3.4.3 MSC Feeding in SFs

To limit depletion of nutrient and growth factors, the replenishment of the concentration of these factors along the culture period is required.

- *Cyclic fed-batch* [24]:

  1. MSC-MC constructs are allowed to settle at the bottom of SFs, by stopping the agitation.

  2. Half of the volume is removed from the SFs and replaced by a same volume of fresh expansion medium or nutrient/growth factors enriched medium.

  3. Continuous agitation is reapplied.

- *Conventional fed batch culture*: Small volumes of culture medium or supplements are continuously added in the SFs during the MSC expansion or differentiation.

3.5 Large-Scale MSC Recovery from Microcarriers

- After expansion or differentiation on MCs, MSCs are enzymatically recovered (i.e., with TrypLE™, Trypsin etc.) in the presence of low mechanical agitation. In the instance where the MCs are not digested by the enzyme, MSCs are separated from the MCs using cell strainers.

- Novel integrated large-scale harvest and separation systems are now available such as HyQ Harvestainer™ BPC (for up to 200 L).

3.6 MSC Differentiation in MC-Based Bioreactors

MSC can be differentiated directly on MCs, which is convenient for further MSC manipulations in vitro or for specific in vivo applications (e.g., transplantation). MSC differentiation on MCs and in SFs may be more efficient than monolayer cultures, potentially due to the mechanical properties of MCs and the hydrodynamic stresses provided by SFs [22].

1. MSC-MC constructs are allowed to settle at the bottom of SFs, by stopping the agitation.

2. The expansion medium is harvested and exchanged with differentiation media.

3. MSCs are incubated with differentiation media for similar time periods as for 2D cultures. The quantification of the differentiation efficiency can be monitored as for 2D cultures (*see* Section 3.2.3) (*see* **Note 11**).

3.7 Up-Scaling MSC Culture in Stirred Bioreactors

Two strategies can be used for increasing the scale of the MSC culture in SFs. Bead-to-bead transfer allows for an increased surface available for cell growth by the addition of fresh beads. Alternatively, MSCs can be reseeded in larger bioreactors containing a larger amount of MCs (*see* **Notes 11** and **12**).

- *Bead-to-bead transfer*: When confluence is reached on MCs, fresh beads are added at a ratio of 1:1. The culture is carried out in expansion medium with intermittent stirring for 4 h (as described above for cell seeding), and on the following days continuous stirring is imposed [21].

- *Large bioreactors*: After the initial phase of expansion in monolayer or in small-scale SFs, MCs are recovered in their entirety and digested with dispase II or TrypLE™ as described above. Then, all MSCs are seeded in a 1-L or higher scale SFs as described above (*see* **Note 13**) [21].

4 Notes

1. Cytodex-3, Cultispher-S, and SoloHill® collagen coated MCs contain porcine gelatin, which facilitates the initial MSCs adhesion. However, these MCs are less suitable for cGMP culture conditions than other types of MCs that are made of xeno-free materials (e.g., Cytopore, Cytodex-1, Hillex II).

2. To be defined as MSCs, the cell population must demonstrate at least 90 % staining for CD90, STRO-1, CD146, CD105, and CD73, while only a negligible percentage of cells (less than 1 %) may be positive for CD14 or CD19.

3. The markers of differentiation are located in the cytoplasm of MSCs. Thus, for their detection, the procedure detailed in Section 3.2.1 can be followed, expecting that cells must be permeabilized using 0.5 % Triton X-100 (or alternative solvent) prior to blocking. To ensure the specific expression of the markers after differentiation, isotype-stained cells must be used as negative controls. The level of expression of differentiation markers must be compared to undifferentiated MSCs cultivated for similar time periods.

4. The endogenous reference gene markers for normalization should be carefully chosen and validated for stability under the experimental conditions. Some widely used reference genes have indeed shown large instability and variation, depending on the MSC source and the culture conditions. Non-validated references can induce significant bias in the analysis of MSC differentiation using RT-qPCR [25].

5. The level of staining for differentiated MSCs must be compared to a negative control, which consists of undifferentiated cells cultivated for the same culture period. The quantitative assessment by spectrophotometry of the MSC differentiation using stainings must be preferred in comparison to qualitative estimation. Alternatively, quantitative large image analysis can be performed (i.e., using ImageJ or Matlab®).

6. CFU-F is a measure of MSC clonogenicity.

7. The seeding efficiency should reach 80–90 % when the above-described procedure is followed (Section 3.3.3).

8. The presence of actin and stress fibers in MSCs seeded on MCs indicates good cell spreading and is suitable for MSC expansion or osteogenic differentiation. In turn, the presence of round cells with absence of actin fibers indicates a suitable adhesion for adipogenic or chondrogenic differentiation [22] (Fig. 1b, reproduced from Sart et al. [13]).

9. The method enables to dynamically monitor changes in growth rate by quantifying the percentage of MSC in S-phase of the cell

cycle. It also allows to define feeding strategies for growth factors (e.g., FGF, TGF-β1), which were demonstrated to be limiting factors in batch culture [24].

10. Measuring dynamically nutrient concentrations in the culture supernatant (e.g., the consumption of glucose and glutamine, the release of lactate and ammonia) is mandatory to design feeding strategies for the replenishment of the concentration of some metabolic components limiting the maximal MSC expansion.

11. In the instance where MSC shape on MCs after expansion is not suitable for certain differentiation pathways (e.g., spread cells on Cultispher-S for chondrogenic differentiation), MSCs can be harvested (*see* Section 3.5) from MCs and reseeded on a different type of MCs more suitable for the differentiation phase (e.g., reseeding on Cytopore-2 for chondrogenic differentiation, which favors more rounded cell morphology) [22]. To ensure the efficient differentiation of MSCs in SFs, the cells must be checked for their phenotype, as described in Section 3.2.3, following the specific indications for MSCs on MCs.

12. *Large-scale expansion and differentiation of MSCs in SFs under cGMP conditions*

To comply with the regulatory guidelines (ATMP in Europe or FDA regulations in the USA), all raw materials, equipment (i.e., MCs, SFs, and all chemicals) and production processes must be implemented according to preestablished standards and specifications [11]. Thus, the culture of MSCs in xeno-free media and with recombinant (non-animal derived) proteins (i.e., for MC coating or for inducing differentiation) is usually preferred for the clinical grade production of MSCs. Under cGMP conditions the potential transmission of pathogens and the potential acquisition of immunogenicity are limited. The present methodology indicates the chemical compounds, the MC types, and the vessels suitable for the cGMP compliant culture and differentiation of MSCs in large scale.

The initial adhesion of MSCs on MC material is usually indirect and is mediated by exogenous ECM binding to the MCs, which is achieved through pre-coating (by proteins from serum, or recombinant proteins). Thus, the homogeneous distribution of coated ECM proteins on the scaffold material should be systematically assessed to ensure the robustness of MSC adhesion on MCs (e.g., by XPS analysis). In addition, the mechanical properties of the substrate to which MSCs adhere play a significant role in the regulation of their fate decision. However, no systematic investigations regarding Young's modulus or Poisson ratio of MCs have been performed so far.

Moreover, it is likely that the MC properties may change during the culture. For instance, the collision between MCs during suspension culture may regulate dynamically the mechanical properties of MCs and, as a consequence, the cell behavior. Thus, investigations on the biochemical and bio-mechanical properties of MCs are mandatory to ensure the reproducibility and the robustness of the culture process in stirred bioreactors.

On the other hand, the culture of MSCs as spheroids has been recently shown to be a promising avenue for the efficient regulation of MSC behavior [26]. Compared to cells attached to scaffolds, these 3D self-organizing structures have demonstrated improved biological performances in terms of differentiation and trophic factor secretion, which are further enhanced when cultivated in SFs [27].

13. *Limitations of the SFs for the control of mechanical induction of MSC fate decision*

As demonstrated with perfusion bioreactors, the controlled delivery of mechanical stresses provides enhanced signaling to promote MSC proliferation and differentiation in 3D cultures. However, while the present methods enable the rapid and convenient large-scale expansion and differentiation of MSCs, recent investigations using computational models (e.g., CFD) have indicated that the hydrodynamics is not well controlled in SFs. Indeed, a broad distribution of fluid velocity and wall shear stress are found within the vessels [28, 29]. Novel vessel and impeller geometry designs are crucially required to better control the fluid motion (e.g., using baffles or wavy walls). Alternatively, the retention of the MCs at a defined location of the culture (e.g., as in basket-based bioreactors) provides more predictable fluid flow stresses to MSCs, to better ensure the reproducibility of their fate decision.

References

1. Linero I, Chaparro O (2014) Paracrine effect of mesenchymal stem cells derived from human adipose tissue in bone regeneration. PLoS One 9:e107001

2. Gao X et al (2014) Bone marrow mesenchymal stem cells promote the repair of islets from diabetic mice through paracrine actions. Mol Cell Endocrinol 388:41–50

3. Wu L et al (2011) Trophic effects of mesenchymal stem cells increase chondrocyte proliferation and matrix formation. Tissue Eng Part A 17:1425–1436

4. Chan JKY, Lam PYP (2013) Human mesenchymal stem cells and their paracrine factors for the treatment of brain tumors. Cancer Gene Ther 20:539–543

5. Crisan M et al (2008) A perivascular origin for mesenchymal stem cells in multiple human organs. Cell Stem Cell 3:301–313

6. Sacchetti B et al (2007) Self-renewing osteoprogenitors in bone marrow sinusoids can organize a hematopoietic microenvironment. Cell 131:324–336

7. Van Dijk CGM et al (2015) The complex mural cell: pericyte function in health and disease. Int J Cardiol 190:75–89

8. Lozito TP et al (2013) Three-dimensional osteochondral microtissue to model pathogenesis of osteoarthritis. Stem Cell Res Ther 4:S6

9. Alexander PG, Gottardi R, Lin H, Lozito TP, Tuan RS (2014) Three-dimensional osteogenic and chondrogenic systems to model

osteochondral physiology and degenerative joint diseases. Exp Biol Med (Maywood) 239:1080–1095

10. Chen AK-L, Reuveny S, Oh SKW (2013) Application of human mesenchymal and pluripotent stem cell microcarrier cultures in cellular therapy: achievements and future direction. Biotechnol Adv 31:1032–1046

11. Sart S, Schneider Y-J, Li Y, Agathos SN (2014) Stem cell bioprocess engineering towards cGMP production and clinical applications. Cytotechnology 66:709–722

12. Van Wezel AL (1967) Growth of cell-strains and primary cells on micro-carriers in homogeneous culture. Nature 216:64–65

13. Sart S, Agathos SN, Li Y (2013) Engineering stem cell fate with biochemical and biomechanical properties of microcarriers. Biotechnol Prog 29:1354–1366

14. Martin Y, Eldardiri M, Lawrence-Watt DJ, Sharpe JR (2011) Microcarriers and their potential in tissue regeneration. Tissue Eng Part B Rev 17:71–80

15. Bertolo A et al (2015) Injectable microcarriers as human mesenchymal stem cell support and their application for cartilage and degenerated intervertebral disc repair. Eur Cell Mater 29:70–80, discussion 80–81

16. Sart S, Ma T, Li Y (2013) Cryopreservation of pluripotent stem cell aggregates in defined protein-free formulation. Biotechnol Prog 29:143–153

17. Baxter MA et al (2004) Study of telomere length reveals rapid aging of human marrow stromal cells following in vitro expansion. Stem Cells Dayt (Ohio) 22:675–682

18. Sepúlveda JC et al (2014) Cell senescence abrogates the therapeutic potential of human mesenchymal stem cells in the lethal endotoxemia model. Stem Cells Dayt (Ohio) 32:1865–1877

19. Dominici M et al (2006) Minimal criteria for defining multipotent mesenchymal stromal cells. The International Society for Cellular Therapy position statement. Cytotherapy 8:315–317

20. Jaiswal N, Haynesworth SE, Caplan AI, Bruder SP (1997) Osteogenic differentiation of purified, culture-expanded human mesenchymal stem cells in vitro. J Cell Biochem 64:295–312

21. Sart S, Schneider Y-J, Agathos SN (2009) Ear mesenchymal stem cells: an efficient adult multipotent cell population fit for rapid and scalable expansion. J Biotechnol 139:291–299

22. Sart S, Errachid A, Schneider Y-J, Agathos SN (2013) Modulation of mesenchymal stem cell actin organization on conventional microcarriers for proliferation and differentiation in stirred bioreactors. J Tissue Eng Regen Med 7:537–551

23. Sart S, Ma T, Li Y (2014) Extracellular matrices decellularized from embryonic stem cells maintained their structure and signaling specificity. Tissue Eng Part A 20:54–66

24. Sart S, Schneider Y-J, Agathos SN (2010) Influence of culture parameters on ear mesenchymal stem cells expanded on microcarriers. J Biotechnol 150:149–160

25. Ragni E, Viganò M, Rebulla P, Giordano R, Lazzari L (2013) What is beyond a qRT-PCR study on mesenchymal stem cell differentiation properties: how to choose the most reliable housekeeping genes. J Cell Mol Med 17:168–180

26. Sart S, Tsai A-C, Li Y, Ma T (2014) Three-dimensional aggregates of mesenchymal stem cells: cellular mechanisms, biological properties, and applications. Tissue Eng Part B Rev 20:365–380

27. Frith JE, Thomson B, Genever PG (2010) Dynamic three-dimensional culture methods enhance mesenchymal stem cell properties and increase therapeutic potential. Tissue Eng Part C Methods 16:735–749

28. Sucosky P, Osorio DF, Brown JB, Neitzel GP (2004) Fluid mechanics of a spinner-flask bioreactor. Biotechnol Bioeng 85:34–46

29. Liovic P, Sutalo ID, Stewart R, Glattauer V, Meagher L (2012) Fluid flow and stresses on microcarriers in spinner flask bioreactors. In: Ninth International Conference on CFD in the Minerals and Process Industries CSIRO, Melbourne, Australia 10–12 December 2012

Methods in Molecular Biology (2016) 1502: 103–110
DOI 10.1007/7651_2016_355
© Springer Science+Business Media New York 2016
Published online: 27 April 2016

Bioreactor Expansion of Skin-Derived Precursor Schwann Cells

Tylor Walsh, Jeff Biernaskie, Rajiv Midha, and Michael S. Kallos

Abstract

Scaling up the production of cells in a culture process is a critical step when trying to develop cell-based regenerative therapies. Static cultures often cannot be easily scaled up to clinically relevant cell numbers. Alternatively, bioreactors offer a highly valuable means to develop a clinical-ready process. To culture adherent cells in suspension, such as skin-derived precursor Schwann cells (SKP-SCs), microcarriers need to be used. Microcarriers are small spherical beads suspended within the vessel that allow for higher growth surface area to volume ratio. Here we describe the procedure of combining microcarriers with the controllability of bioreactors to generate higher cell densities in smaller reactor volumes leading to a more efficient and cost-effective cell production for applications in regenerative medicine.

Keywords: Bioreactors, Scale-up, Schwann cells, SKP-SCs, Microcarriers, Regenerative medicine

1 Introduction

Peripheral nervous system (PNS) axons are ensheathed by Schwann cells forming myelin. Following injury, PNS axons exhibit a limited ability to repair itself, in which Schwann cells play a major supporting role [1]. Depending on the location and extent of the nerve injury, it can result in sensory loss or paralysis and is frequently associated with pain [2]. The two main current treatments are tensionless epineural sutures or repair of a nerve gap with autologous nerve grafting [2]. Clinical treatments for nerve injuries have hardly changed in 30 years and result in only modest recovery [2]. Because of this, new treatments are being investigated with much effort towards developing cellular therapies and regenerative treatments that exploit endogenous mechanisms of repair. Our group has focused on using skin-derived precursor cells (SKPs) which can be readily differentiated into Schwann cells, referred to as skin derived precursor Schwann cells (SKP-SCs) [3–6]. SKP-SCs are easily accessible, allow for an autologous source of cells for treatments, and show great potential to promote axonal regeneration, generate new myelin and encourage reinnervation of damaged nerves. However, processes for efficiently expanding SKP-SCs remain suboptimal as current static culture methods are not adequate to produce the high numbers of cells needed for timely treatments. Moreover, in static cultures, most culture variables are not controlled, leading to large batch-to-batch variations as controlling, monitoring, and evaluating the impact of key parameters on target cell output and productivity is difficult [7]. Because of this, bioreactors are an attractive option. Bioreactors, in the context of cell culture, are vessels or devices that allow the user to regulate environmental conditions the cells experience by controlling factors such as dissolved oxygen, pH, nutrients, temperature, and shear stress [8]. Given their ability to control the culture environment, bioreactors have been used in a number of applications including culture of primary tissues and derivatives with intent to repair damaged tissue, proliferate and differentiate stem cells to desired cell types, and "scale-out" or "scale-up" operations to obtain enough cells for autologous or allogeneic clinical therapies, respectively [8]. A number of cell types can be cultured in bioreactors, either as single cells in suspension, as aggregates, or attached to microcarriers [9]. SKP-SCs are an adherent cell type, and as such, they require microcarriers in order to grow in suspension bioreactors. Microcarriers are small beads with diameters of 100–400 μm, which can be manufactured out of a number a materials [10]. They are typically formed of a stable matrix that can withstand the mechanical forces seen in a bioreactor. When inoculated, the cells will attach to the microcarriers and proliferate. However, this attachment depends on many factors including chemical composition, surface topography, degree of porosity, and charge density of the particles [10, 11]. The large surface area to volume ratio that the microcarriers create allows for the culturing of many cells in a small volume. Additionally, serial sampling can be done without causing disruption of cell proliferation and regulation of differentiation can be achieved through the addition of appropriate stimuli [10]. We have been able to successfully culture the SKP-SCs on microcarriers in uncontrolled spinner flask bioreactors [12]. By moving to computer controlled bioreactors, we will be able to obtain higher maximum cell densities showing that these cells have great potential for scale up.

2 Materials

2.1 Medium Preparation

1. DMEM, low glucose with L-glutamine and sodium pyruvate (Life Technologies).

2. F12 (1×) with GlutaMAX (Life Technologies).

3. Neuregulin (recombinant human E. coli derived rh NRG-1-β1-Neuregulin, Fisher Scientific).

4. Forskolin (Sigma-Aldrich).

5. Penicillin–streptomycin (Life Technologies).

6. N2 Supplement (Life Technologies).

7. Fetal bovine serum, FBS (Life Technologies).

2.2 Bioreactor Culture

1. Cytodex 3 microcarriers (GE Healthcare).

2. Phosphate-buffered saline, PBS (Life Technologies).

3. Crystal violet (Sigma-Aldrich).

4. Methanol (VWR).

5. Citric acid (Sigma-Aldrich).

6. Double distilled water (DDW).

7. DASGIP parallel bioreactor system (Eppendorf).

8. TrypLE Express (Life Technologies).

9. Trypan Blue solution (Sigma-Aldrich).

10. Erlenmeyer flasks.

11. Conical tube 70 μm cell strainer (BD Falcon).

12. Bottle top filter (Thermo Scientific).

13. Sigmacote (Sigma-Aldrich).

14. Aluminum foil.

15. 50 and 15 mL conical tubes (FroggaBio).

16. 70 % Ethanol in DDW.

17. Hemocytometer (Hausser Scientific).

18. Antibiotic-Antimycotic (Anti-Anti, Life Technologies).

3 Methods

3.1 SKP-SC Medium

1. Add DMEM and F12 in a ratio of 3:1 with 1 % penicillin–streptomycin, 5.0 μM forskolin, 50.0 ng/mL neuregulin, 1 % N2 supplement, and 1 % FBS (*see* **Note 1**).

2. Filter through 0.2 μm bottle top filter (*see* **Note 2**).

3.2 Siliconization of Flasks and Reactors

1. Determine how many Erlenmeyer flasks are needed to prepare the microcarriers and how many reactors are needed for the experiment and move them to a fume hood (*see* **Note 3**).

2. Add 10.0 mL of Sigmacote to bottom of flasks or reactors and rotate to coat walls above working volume liquid level (*see* **Note 4**).

3. Coat impellers of the reactors including the blades and up the shaft.

4. Remove Sigmacote and place back in stock bottle.

5. Let the reactors and flasks dry for 8–24 h in the fume hood.

6. Rinse flasks and reactors with Ca^{2+}/Mg^{2+} free PBS; then fill to working volume with PBS and let sit for 8–24 h.

7. Rinse reactors and flasks with double distilled water (DDW) and fill to working volume with DDW and let sit 8–24 h.

8. Remove water and let flasks/reactors dry thoroughly and store until needed.

3.3 Microcarrier Preparation

3.3.1 Hydration

1. Weigh desired amount of Cytodex 3 microcarriers dry and transfer to a siliconized 125 mL Pyrex flask and hydrate by adding 100.0 mL of Ca^{2+}/Mg^{2+} free PBS with 1 % Anti-Anti. The microcarriers are swollen and hydrated in PBS overnight at room temperature (*see* **Note 5**).

2. After 24 h, remove the supernatant using a 10 mL pipette and wash the microcarriers in 100.0 mL of fresh Ca^{2+}/Mg^{2+} free PBS with 1 % Anti-Anti twice, letting the microcarriers settle between washes.

3. Add 100.0 mL of fresh Ca^{2+}/Mg^{2+} free PBS with 1 % Anti-Anti to the flask and store at 4 °C overnight or until needed (*see* **Note 6**).

3.3.2 Sterilization

1. Discard PBS using a 10 mL pipette reducing the volume of settled microcarriers to approximately 20.0 mL, and replace with 30.0 mL of fresh PBS with 1 % Anti-Anti for a total volume of 50.0 mL.

2. Place aluminum foil over top of flask and sterilize by autoclaving at 120 °C and 152 kPa (23.25 psi) for 60 min.

3. Microcarriers are now ready to use or can be sealed with Parafilm and place at 4 °C for later use (*see* **Note 6**).

3.4 Bioreactor Set-Up

1. Prepare reactors by siliconizing vessels, impellers and all probes as described above.

2. Set up control program by setting the dissolved oxygen, pH, agitation rate, inlet gas flow, liquid pump rate, and temperature (*see* **Note 7**).

3. Connect pH probe and perform a two-point calibration at pH 4 and pH 7.

4. Wrap tops of probes and bioreactors in aluminum foil, add 1.0 mL of DDW to the reactors, and sterilize by autoclave on fluid cycle (*see* **Note 8**).

5. After sterilization, move reactors to biosafety cabinet and let cool. Next add microcarriers and 50 % of final volume of SKP-SC medium to reactors and connect all probes (pH, DO, temperature, heating jacket, and gas inlet) (*see* **Note 9**).

6. Perform a two-point DO calibration by first disconnecting the DO probe and calibrating 0 %, then reconnect the probe and

set oxygen at 21 % and wait a minimum of 6 h and calibrate the 100 % point.

7. Disconnect all probes, move reactors to biosafety cabinet, and inoculate cells.

3.5 Bioreactor Inoculation and Culture

1. Passage cells from static culture (10 cm dish) when they are 80 % confluent.

2. Remove medium and wash twice with 5 mL of DMEM.

3. Add 3.0 mL of TrypLE Express and incubate at 37 °C for 5 min.

4. Remove cells and place into conical tube.

5. Rinse dish twice with 5.0 mL DMEM and place into conical tube.

6. Centrifuge at 300 × g, remove supernatant, and resuspend in 5.0 mL SKP-SC medium (*see* **Note 2**).

7. Count cells using trypan blue exclusion method and inoculate cells at desired concentration of cells/microcarrier (*see* **Note 10**).

8. To inoculate cells, disconnect probes and take reactors to biosafety cabinet. Add cells and SKP-SC Medium to 60 % of final volume. Reconnect reactors to control system (*see* **Note 11**).

9. After 24 h, disconnect the probes and move reactors to biosafety cabinet and add remaining SKP-SC medium to final volume of 500.0 mL (or desired volume) and reconnect to control system.

3.6 Bioreactor Sampling

3.6.1 Sampling for Counting

1. Spray bioreactor sampling port and a 5 mL syringe with 70 % ethanol.

2. Connect the syringe on the sampling port, draw 3.0 mL, disconnect the syringe and place sample into 15 mL conical tube.

3. Spray the sampling port with 70 % ethanol.

4. Let microcarriers settle in conical tube and remove supernatant.

5. Wash twice with 1.0 mL PBS, letting microcarriers settle for 3 min between washes.

6. Add 1.0 mL of 0.1 % crystal violet in 0.1 M citric acid and incubate at 37 °C for at least 1 h (*see* **Note 12**).

7. Titrate sample 30 times with a 1 mL pipette to release nuclei, dilute with PBS if necessary, and count using a hemocytometer.

3.6.2 Sampling for Photomicrographs

1. Add 1.5 mL of PBS to a single well of a 6-well plate.

2. Spray bioreactor sampling port and a 5 mL syringe with 70 % ethanol.

3. Connect the syringe on the sampling port, draw 0.5 mL, disconnect the syringe and place sample into the well on the 6-well plate.

4. Spray the sampling port with 70 % ethanol.

5. Add 20.0 µL of 0.5 % crystal violet in methanol and let sit for 5 min at room temperature (*see* **Note 12**).

6. Take photomicrographs (*see* **Note 13**).

3.7 Harvesting Cells from Bioreactors

1. Disconnect reactors and move to biosafety cabinet.

2. Let microcarriers settle for 5 min and remove supernatant.

3. Wash microcarriers twice with 50.0 mL of PBS letting the microcarrier settle for 5 min between washes.

4. Add 100.0 mL of TrypLE Express to reactor and incubate at 37 °C and 100 rpm for 15 min.

5. Let microcarriers settle, remove supernatant, and strain through a 70 µm strainer into conical tubes.

6. Wash twice with 100.0 mL of PBS, letting microcarriers settle and placing supernatant through the strainer into conical tubes.

7. Add 100.0 mL of PBS to the microcarriers and add microcarriers and PBS to strainer.

8. Centrifuge conical tubes at 1000 rpm for 5 min.

9. Remove supernatant, resuspend in SKP-SC medium and count cells using trypan blue method.

4 Notes

1. We find it is best to make the medium fresh the day that it is needed.

2. It is important to warm up the medium in the incubator, with the lid cracked, prior to using to ensure it is at the proper pH and temperature.

3. The number of reactors needed is based on experimental design. The number of flasks needed is determined based on the amount of microcarriers needed. To get efficient hydration of microcarriers do not exceed 1.0 g/flask which gives a concentration of 1.0 g/100.0 mL.

4. Tilt the flask or reactor at an angle and rotate the vessels at least three full turns coating the sides up to the 125 mL mark on the flask and up to the arms on the reactor vessels. For the reactor vessels it is important to pay particular attention to coating the bump at the bottom of the vessel.

5. If the microcarriers float on the surface during the hydration step, one or two drops of a surfactant such as Tween 80 (United State Chemical Corporation) can be added to reduce the surface tension and allow the microcarriers to settle.

6. Microcarriers can be stored for up to a week following hydration or sterilization.

7. DASGIP Parallel bioreactor system process parameters for SKP-SCs : DO was set at 21 %, pH at 7.4, agitation at 40 rpm, inlet gas flow at 3 L/h, liquid pump rate at 0 mL/h, and temperature at 37 °C.

8. It is important to wrap the connections for DO and pH probes individually with aluminum foil. Also wrap the filters to prevent water from getting in. Next wrap the caps of the arms and then wrap the entire top of the reactor and any hosing that is present. It is important to loosen the caps on the reactors to prevent pressure build up during autoclaving.

9. After sterilization, transfer the flask of microcarriers to the biosafety cabinet. Remove the supernatant and rinse with warm DMEM. Make sure that the microcarriers stay in suspension by mixing with a pipette. Using a 10 mL pipette transfer 5.0 mL of microcarrier solution from the flask to a 50 mL conical tube. Repeat this until all the microcarriers are in the conical tube. Let the microcarriers settle then remove the supernatant. Resuspend the microcarriers in 25.0 mL SKP-SC medium and add to bioreactor 5 mL at a time using a 10 mL pipette. For the SKP-SCs we added enough microcarriers for a final concentration or 2 g/L.

10. Optimal inoculation density has to be determined for each cell type. We found that 4 cells/microcarrier works well for the SKP-SCs.

11. At this point you can add the entire volume of medium to a total of 500.0 mL (or desired final volume), but we have found that by reducing the volume to 60 % for the first 24 h increases the cell to microcarrier attachment efficiency.

12. Make sure to wear proper personal protective equipment (lab coat, gloves, and mask) when making crystal violet solutions.

13. To obtain the best photos, take them at 10× or 20× magnification while leaving the filter on phase 1. This will give a flatter image that refracts less light making it the cells easier to see.

Acknowledgments

This study was supported by a CRIO Project grant from Alberta Innovates-Health Solutions (AIHS), Project Grant #20140910 to Rajiv Midha, Michael Kallos, and Jeff Biernaskie.

References

1. Oudega M, Xu XM (2006) Schwann cell transplantation for repair of the adult spinal cord. J Neurotrauma 23:453–467

2. Faroni A, Mobasseri SA, Kingham PJ, Reid AJ (2014) Peripheral nerve regeneration: experimental strategies and future perspectives. Adv Drug Deliv Rev 82–83:160–167

3. Biernaskie J, Sparling JS, Liu J, Shannon CP, Plemel JR, Xie Y, Miller FD, Tetzlaff W (2007) Skin-derived precursors generate myelinating Schwann cells that promote remyelination and functional recovery after contusion spinal cord injury. J Neurosci 27:9545–9559

4. Khuong HT, Kumar R, Senjaya F, Grochmal J, Ivanovic A, Shakhbazau A, Forden J, Webb A, Biernaskie J, Midha R (2014) Skin derived precursor Schwann cells improve behavioral recovery for acute and delayed nerve repair. Exp Neurol 254:168–179

5. McKenzie IA, Biernaskie J, Toma JG, Midha R, Miller FD (2006) Skin-derived precursors generate myelinating Schwann cells for the injured and dysmyelinated nervous system. J Neurosci 26:6651–6660

6. Biernaskie J, McKenzie IA, Toma JG, Miller FD (2006) Isolation of skin-derived precursors (SKPs) and differentiation and enrichment of their Schwann cell progeny. Nat Protoc 1:2803–2812

7. Kirouac DC, Zandstra PW (2008) The systematic production of cells for cell therapies. Cell Stem Cell 3:369–381

8. Naing MW, Williams DJ (2011) Three-dimensional culture and bioreactors for cellular therapies. Cytotherapy 13:391–399

9. Yuan Y, Kallos MS, Hunter C, Sen A (2012) Improved expansion of human bone marrow-derived mesenchymal stem cells in microcarrier-based suspension culture. J Tissue Eng Regen Med 8:210–225

10. Martin Y, Eldardiri M, Lawrence-Watt DJ, Sharpe JR (2011) Microcarriers and their potential in tissue regeneration. Tissue Eng Part B Rev 17:71–80

11. Chen AKL, Chen X, Choo ABH, Reuveny S, Oh SKW (2011) Critical microcarrier properties affecting the expansion of undifferentiated human embryonic stem cells. Stem Cell Res 7:97–111

12. Shakhbazau A, Mirfeizi L, Walsh T, Wobma HM, Kumar R, Singh B, Kallos MS, Midha R (2016) Inter-microcarrier transfer and phenotypic stability of stem cell-derived Schwann cells in stirred suspension bioreactor culture. Biotechnol Bioeng 113:393–402

Methods in Molecular Biology (2016) 1502: 111–118
DOI 10.1007/7651_2016_334
© Springer Science+Business Media New York 2016
Published online: 12 April 2016

Use of Stirred Suspension Bioreactors for Male Germ Cell Enrichment

Sadman Sakib*, Camila Dores*, Derrick Rancourt, and Ina Dobrinski

Abstract

Spermatogenesis is a stem cell based system. Both therapeutic and biomedical research applications of spermatogonial stem cells require a large number of cells. However, there are only few germ line stem cells in the testis, contained in the fraction of undifferentiated spermatogonia. The lack of specific markers makes it difficult to isolate these cells. The long term maintenance and proliferation of nonrodent germ cells in culture has so far been met with limited success, partially due to the lack of highly enriched starting populations. Differential plating, which depends on the differential adhesion properties of testicular somatic and germ cells to tissue culture dishes, has been the method of choice for germ cell enrichment, especially for nonrodent germ cells. However, for large animals, this process becomes labor intensive and increases variability due to the need for extensive handling. Here, we describe the use of stirred suspension bioreactors, as a novel system for enriching undifferentiated germ cells from 1-week-old pigs. This method capitalizes on the adherent properties of somatic cells within a controlled environment, thus promoting the enrichment of progenitor cells with minimal handling and variability.

Keywords: Germ cells, Spermatogonial stem cell, Testes, Stirred suspension bioreactor, Spinner flasks

1   Introduction

Male germ line stem cells, located at the basement membrane of seminiferous tubules, form the foundation of spermatogenesis. These cells have widespread potential applications: for treating male infertility, preserving endangered species, regenerative medicine, and generating transgenic animals [1–4]. Germ line stem cells, once isolated from the testis and placed into a conducive environment, have the ability to attain pluripotency and differentiate into tissues of all three germ lineages [5–11]. However, a significant number of cells are required for these applications.

Expansion of germ cell number is a challenge due to the lack of efficient enrichment protocols, especially for nonrodent model species such as pig. The current enrichment method being employed is differential plating, which is well established for rodent

*These authors contributed equally to this work

germ cells [12]. The differential plating technique depends on the differential adhesion of germ and somatic cells to tissue culture surfaces. It involves the sequential plating of testicular cell suspensions, where Sertoli cells readily adhere to the plastic surfaces of tissue culture plates while germ cells remain suspended in the supernatant or attach slightly to the somatic cells. In addition to collecting germ cells from the supernatant, those in the slightly attached cell fraction can be obtained by treatment with 1:20 and 1:5 dilution of trypsin–EDTA 0.25 % .

Although this technique has been employed with satisfactory results [13, 14], when dealing with large animal models, this technique becomes too labor intensive due to the large number of cells involved. It also increases the risk of introducing variability due to extensive cell handling. Therefore, a protocol that allows efficient enrichment with minimum handling within a controlled environment is needed.

Stirred suspension bioreactors provide an efficient system for the expansion of cells in a controlled environment, reducing sources of variability. Bioreactors have been employed in the culture and long term maintenance of a number of different cell types [15–17]. Here, we report two protocols for enriching prepubertal (1 week old) porcine germ cells by exploiting the adhesive of Sertoli cells in bioreactors [18]. One protocol involves using the bioreactor solely. The starting cell suspension is stirred at 100 rpm for 66 h allowing Sertoli cells to form aggregates. At 18, 44, and 66 h, cell suspensions are filtered through a 40 μm mesh to remove the aggregates, resulting in a seven to eightfold enrichment of germ cells. The second protocol involves two rounds of bioreactor filtration as above (44 h) followed by differential plating. Cell suspensions were stirred at 100 rpm for 44 h. At 18 and 44 h the cells were filtered to remove the Sertoli cell aggregates. After the second round of filtration, the remaining cells were trypsinized and plated for 22 h. The supernatant and the slightly adhered fraction of germ cells were collected at 66 h. This approach consistently resulted in a 9-fold enrichment of germ cells [18]. A schematic of the experimental design is shown in Fig. 1.

2   Materials

All solutions must be prepared under sterile conditions and reagents tested for cell culture must be used.

2.1   Testes Tissue Enzymatic Digestion Reagents

1. DPBS (w/o Ca^{++} and Mg^{++}): Dulbecco's phosphate-buffered saline (w/o Ca^{++} and Mg^{++}), $1\times$ penicillin/streptomycin (p/s).

2. HDMEM $1\times$ p/s: High glucose Dulbecco's modified Eagle medium (HDMEM), $1\times$ p/s.

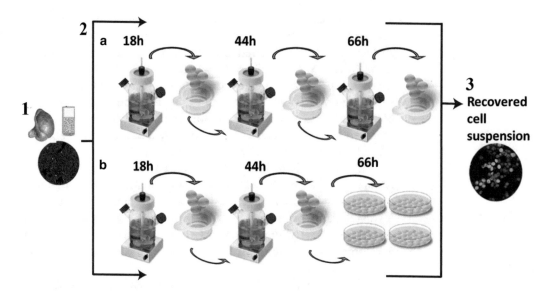

Fig. 1 Experimental design. *1*. 1-week-old pig testes are enzymatically digested into a single cell suspension. *2*. Cell suspensions were split into two groups and assigned to enrichment method A and B. *2A*. Cell suspensions were placed in stirred suspension culture, after 18, 44, and 66 h cell suspensions were filtered through a 40 μm strainer to remove larger cell aggregates and poured back into culture until the next time point. *2B*. Cell suspensions were placed in stirred suspension culture for 44 h, then subjected to one round of differential plating. *3*. Enriched germ cell suspension is obtained. Figure is modified from [18] and reproduced with permission

3. HDMEM: High glucose Dulbecco's Modified Eagle medium (HDMEM) without additives.

4. Collagenase IV: HDMEM, 2 mg/mL collagenase IV (*see* **Note 1**).

5. Hyaluronidase: HDMEM, 2.5 μg/mL hyaluronidase (*see* **Note 2**).

6. DNase I: HDMEM, 7 mg/mL DNase I (*see* **Note 3**).

7. 0.25 % trypsin–EDTA.

8. Culture dish: 100 mm tissue culture dishes.

9. Sterile forceps, scissors.

2.2 Stirred Suspension Bioreactor Culture

1. Bioreactor vessel: 125 mL Stirred Suspension Bioreactors (Corning Style Spinner Flask; NDS Technologies Inc. Vineland, NJ, USA) with magnetic impellers.

2. Magnetic stir plate: Variable speed magnetic stir plates that can maintain a speed of 100 rpm and can be housed within a CO_2 cell culture apparatus.

3. Bioreactor culture medium (HDMEM): High-glucose Dulbecco's modified Eagle medium (DMEM), 50 U/mL penicillin, 50 U/mL streptomycin, 5 % fetal bovine serum (FBS).

4. DPBS (w/o Ca^{++} and Mg^{++}): Dulbecco's phosphate-buffered saline DPBS (w/o Ca^{++} and Mg^{++}).

5. 0.25 % trypsin–EDTA.

2.3 Stirred Suspension Bioreactor Followed by Differential Plating

1. Bioreactor culture medium (HDMEM): High-glucose Dulbecco's modified Eagle medium (DMEM), 50 U/mL penicillin, 50 U/mL streptomycin, 5 % fetal bovine serum (FBS).

2. 0.25 % trypsin–EDTA.

3. 0.25 % trypsin–EDTA (1:10): 36 mL DPBS, 4 mL 0.25 % trypsin–EDTA.

4. Culture dish: 100 mm tissue culture dishes.

5. DPBS (w/o Ca^{++} and Mg^{++}): Dulbecco's phosphate-buffered saline DPBS (w/o Ca^{++} and Mg^{++}).

3 Methods

3.1 Testes Tissue Enzymatic Digestion

1. The testes are obtained by castration of 1 week old pigs (*see* **Note 4**).

   The isolated porcine testes are washed in 3 changes of DPBS 1× p/s at room temperature (*see* **Note 5**).

2. Using sterile forceps the tunica vaginalis and epididymis are cut and the testis tissue is placed in a new clean dish.

3. The dish is closed and tissue is weighed together (*see* **Note 6**).

4. Using sterile forceps and scissors, the testis capsule is peeled from the parenchyma and the capsule is placed in a clean dish.

5. The capsule and the dish are weighed together and subtracted from the initial weight to obtain the actual weight of the testes tissue being digested.

6. The tissue is cut in half and then in half again lengthwise and the mediastinum is removed (*see* **Note 7**).

7. The remaining tissue is minced with sterile scissors into ~1–2 mm size pieces and placed into 40 mL of collagenase IV (2 mg/mL) using a spatula (*see* **Note 8**).

8. The tubes are then placed in 37 °C water bath and inverted every 5 min until the tissue appears more uniform and the fluid becomes cloudy (~30 min).

9. The tissue is allowed to sediment and the supernatant is discarded.

10. Then, 20 mL collagenase IV (2 mg/mL), 20 mL hyaluronidase (2.5 μg/mL) and 1 mL DNase I (7 mg/mL) are added to the tissue.

11. The tubes are placed back in 37 °C water bath and inverted every 5 min for ~30 min.

12. When almost no tissue fragments remain and seminiferous cords are floating, the tissue is centrifuged at $75 \times g$ for 5 min.

13. The supernatant is removed and the tubules are resuspended in DPBS and transferred to a new 50 mL tube leaving the chunks behind.

14. The tubules are then centrifuged at $75 \times g$ for 2 min and washed with DPBS thrice.

15. The supernatant from the last wash is removed and 5 mL of H-DMEM, $1 \times$ p/s is added.

16. 10 mL of 0.25 % trypsin–EDTA is added and mixed by inverting and placed in the 37 °C water bath.

17. After 5 min the tube is inverted and observed for the presence of cloudy white DNA. If DNA is present, 500–1000 µL of DNase I is added to the tube.

18. The tubes are continuously shaken and observed every 5 min; if necessary trypsin and DNase maybe added (*see* **Note 9**).

19. Upon satisfactory digestion, 10 % FBS is added to neutralize the trypsin.

20. The cell suspension is then filtered through 100, 70 and then 40 µm cell strainers.

21. The tubes are then centrifuged at $500 \times g$ for 5 min.

22. The supernatant is discarded and the cells are resuspended in 30–40 mL fresh H-DMEM, which is considered here as the starting cell suspension.

3.2 Stirred Suspension Bioreactor Culture

1. An appropriate volume of starting cell suspension containing 500×10^6 cells is introduced into the bioreactor.

2. Pre-warmed high-glucose Dulbecco's modified Eagle medium (HDMEM) supplemented with 50 U/mL penicillin, 50 U/mL streptomycin, and 5 % fetal bovine serum (FBS) is added to make a volume of 100 mL.

3. The bioreactor is then placed on magnetic stirrer in the cell culture incubator and the cells are stirred at 100 rpm in 37 °C and 5 % CO_2 (*see* **Notes 10** and **11**).

4. After 18 and 44 h, all the media and the cells are removed from the bioreactor; filtered through a 40 µm mesh to remove large cell aggregates and placed into two sterile 50 mL centrifuge tubes (*see* **Note 12**).

5. The two sterile 50 mL centrifuge tubes are then centrifuged at $500 \times g$ for 5 min and the supernatant is discarded.

6. The sedimented cells are then resuspended in fresh pre-warmed HDMEM.

7. The suspension is again placed in the bioreactor and cultured as described in **step 3** of this section.

8. After 66 h from the initial loading, the cells are again filtered through a 40 μm filter and concentrated as cell pellets after 5 min centrifugation at $500 \times g$.

9. The cell pellet is then resuspended in 0.25 % trypsin–EDTA for 5 min to obtain a single cell suspension from the remaining cells by breaking up aggregates smaller than 40 μm (*see* **Note 13**).

10. The trypsin is neutralized by addition of 10 % FBS and the single cell suspension is centrifuged at $500 \times g$ for 5 min.

11. The supernatant is then discarded and the cells are washed twice in PBS and resuspended in HDMEM.

3.3 Stirred Suspension Bioreactor Followed by Differential Plating

1. Testicular cells (5×10^6 cells/mL) will be subjected to enrichment by SSB as described above in Sect. 3.1 for 44 h. The recovered cells are centrifuged at $500 \times g$ for 5 min and concentrated as pellets.

2. The pellet is then resuspended in 0.25 % trypsin–EDTA (5 min) until a single cell suspension is observed. The trypsin is neutralized by the addition of 10 % FBS, the cells are washed with PBS twice and resuspended in high-glucose DMEM with 5 % FBS (*see* **Note 13**).

3. The cells are seeded at 5×10^6 cells/mL concentration on 100 mm tissue culture dishes in 10 mL of high-glucose Dulbecco's modified Eagle medium (HDMEM) and incubated for 22 h at 37 °C in 5 % CO_2 in air.

4. After 22 h (66 h since the initial loading), the cells in suspension and the slightly attached fractions are collected by trypsinization for 30 s with 1:10 dilution of 0.25 % trypsin–EDTA.

5. The trypsin is then neutralized by adding 10 % FBS and the cells are centrifuged at $500 \times g$ for 5 min.

6. The supernatant is removed and the pellet is washed twice with DPBS.

7. The cells are then resuspended in high-glucose Dulbecco's modified Eagle medium (HDMEM).

4   Notes

1. Dissolve 80 mg of Collagenase IV powder in 40 mL H-DMEM w/o additives in a 50 mL centrifuge tube and filter through a 0.2 μm vacuum filter.

2. Dissolve 0.05 mg Hyaluronidase in 20 mL of HDMEM w/o additives and filter through a 0. 2 μm vacuum filter.

3. Resuspend 56 mg of DNase I in 8 mL H-DMEM in a 15 mL centrifuge tube and filter through a 0.2 μm syringe filter.

4. This protocol was established for germ cells from prepubertal (1 week old) testes. Germ cells from older animals may result in lower enrichment.

5. DPBS $1\times$ P/S is only required for these initial three washes.

6. If the weighing apparatus is outside of the biosafety cabinet, take care to keep the dish closed to avoid contaminating the tissue.

7. 1–2 cm pieces of tissue may be cut and fixed for subsequent histology.

8. 8 mL (less than 10 mL) of tissue per tube should be added.

9. The amount of cord digestion may be visualized using a microscope. If digestion is going slowly, mechanical digestion by pipetting up and down may be considered.

10. It is recommended to count the rotation of the magnetic bar per minute rather than depending solely on the display of the magnetic stir plate.

11. Higher rpm result in smaller aggregate size, whereas lower rpm, e.g., 80 rpm will result in aggregates of larger size. Different cell types may require different rotation speeds for satisfactory enrichment.

12. The aggregates may be preserved for fixing and staining later to determine the ratio of somatic and germ cells in the aggregates, as some germ cells will be adhered to the somatic cell clumps.

13. In some instances, trypsinization can result in cell degradation and release of the cell DNA. If the presence of DNA (appearance of cloud like structures) is observed, some DNase can be added.

References

1. Sehgal L, Usmani A, Dalal SN, Majumdar SS (2014) Generation of transgenic mice by exploiting spermatogonial stem cells in vivo. Methods Mol Biol 1194:327–337. doi:10.1007/978-1-4939-1215-5_18

2. Wu Y, Zhou H, Fan X, Zhang Y, Zhang M, Wang Y, Xie Z, Bai M, Yin Q, Liang D, Tang W, Liao J, Zhou C, Liu W, Zhu P, Guo H, Pan H, Wu C, Shi H, Wu L, Tang F, Li J (2015) Correction of a genetic disease by CRISPR-Cas9-mediated gene editing in mouse spermatogonial stem cells. Cell Res 25(1):67–79. doi:10.1038/cr.2014.160

3. Sato T, Sakuma T, Yokonishi T, Katagiri K, Kamimura S, Ogonuki N, Ogura A, Yamamoto T, Ogawa T (2015) Genome editing in mouse spermatogonial stem cell lines using TALEN and double-nicking CRISPR/Cas9. Stem Cell Rep 5(1):75–82. doi:10.1016/j.stemcr.2015.05.011

4. Aponte PM (2015) Spermatogonial stem cells: current biotechnological advances in reproduction and regenerative medicine. World J Stem Cells 7(4):669–680. doi:10.4252/wjsc.v7.i4.669

5. Kanatsu-Shinohara M, Lee J, Inoue K, Ogonuki N, Miki H, Toyokuni S, Ikawa M, Nakamura T, Ogura A, Shinohara T (2008) Pluripotency of a single spermatogonial stem cell in mice. Biol Reprod 78(4):681–687. doi:10.1095/biolreprod.107.066068

6. Simon L, Ekman GC, Kostereva N, Zhang Z, Hess RA, Hofmann MC, Cooke PS (2009) Direct transdifferentiation of stem/progenitor spermatogonia into reproductive and nonreproductive tissues of all germ layers.

Stem Cells 27(7):1666–1675. doi:10.1002/stem.93

7. Ning L, Goossens E, Geens M, Van Saen D, Van Riet I, He D, Tournaye H (2010) Mouse spermatogonial stem cells obtain morphologic and functional characteristics of hematopoietic cells in vivo. Hum Reprod 25(12):3101–3109. doi:10.1093/humrep/deq269

8. Kossack N, Meneses J, Shefi S, Nguyen HN, Chavez S, Nicholas C, Gromoll J, Turek PJ, Reijo-Pera RA (2009) Isolation and characterization of pluripotent human spermatogonial stem cell-derived cells. Stem Cells 27(1):138–149. doi:10.1634/stemcells.2008-0439

9. Dym M, He Z, Jiang J, Pant D, Kokkinaki M (2009) Spermatogonial stem cells: unlimited potential. Reprod Fertil Dev 21(1):15–21

10. Golestaneh N, Kokkinaki M, Pant D, Jiang J, DeStefano D, Fernandez-Bueno C, Rone JD, Haddad BR, Gallicano GI, Dym M (2009) Pluripotent stem cells derived from adult human testes. Stem Cells Dev 18(8):1115–1126. doi:10.1089/scd.2008.0347

11. Izadyar F, Wong J, Maki C, Pacchiarotti J, Ramos T, Howerton K, Yuen C, Greilach S, Zhao HH, Chow M, Chow YC, Rao J, Barritt J, Bar-Chama N, Copperman A (2011) Identification and characterization of repopulating spermatogonial stem cells from the adult human testis. Hum Reprod 26(6):1296–1306. doi:10.1093/humrep/der026

12. Hamra FK, Gatlin J, Chapman KM, Grellhesl DM, Garcia JV, Hammer RE, Garbers DL (2002) Production of transgenic rats by lentiviral transduction of male germ-line stem cells. Proc Natl Acad Sci U S A 99(23):14931–14936. doi:10.1073/pnas.222561399

13. Hofmann MC, Braydich-Stolle L, Dym M (2005) Isolation of male germ-line stem cells; influence of GDNF. Dev Biol 279(1):114–124. doi:10.1016/j.ydbio.2004.12.006

14. Izadyar F, Spierenberg GT, Creemers LB, den Ouden K, de Rooij DG (2002) Isolation and purification of type A spermatogonia from the bovine testis. Reproduction 124(1):85–94

15. Shafa M, Sjonnesen K, Yamashita A, Liu S, Michalak M, Kallos MS, Rancourt DE (2012) Expansion and long-term maintenance of induced pluripotent stem cells in stirred suspension bioreactors. J Tissue Eng Regen Med 6(6):462–472. doi:10.1002/term.450

16. Krawetz R, Taiani JT, Liu S, Meng G, Li X, Kallos MS, Rancourt DE (2010) Large-scale expansion of pluripotent human embryonic stem cells in stirred-suspension bioreactors. Tissue Eng Part C Methods 16(4):573–582. doi:10.1089/ten.TEC.2009.0228

17. Fluri DA, Tonge PD, Song H, Baptista RP, Shakiba N, Shukla S, Clarke G, Nagy A, Zandstra PW (2012) Derivation, expansion and differentiation of induced pluripotent stem cells in continuous suspension cultures. Nat Methods 9(5):509–516. doi:10.1038/nmeth.1939

18. Dores C, Rancourt D, Dobrinski I (2015) Stirred suspension bioreactors as a novel method to enrich germ cells from pre-pubertal pig testis. Andrology 3(3):590–597. doi:10.1111/andr.12031

Methods in Molecular Biology (2016) 1502: 119–128
DOI 10.1007/7651_2015_310
© Springer Science+Business Media New York 2015
Published online: 03 February 2016

Generation of Neural Progenitor Spheres from Human Pluripotent Stem Cells in a Suspension Bioreactor

Yuanwei Yan, Liqing Song, Ang-Chen Tsai, Teng Ma, and Yan Li

Abstract

Conventional two-dimensional (2-D) culture systems cannot provide large numbers of human pluripotent stem cells (hPSCs) and their derivatives that are demanded for commercial and clinical applications in in vitro drug screening, disease modeling, and potentially cell therapy. The technologies that support three-dimensional (3-D) suspension culture, such as a stirred bioreactor, are generally considered as promising approaches to produce the required cells. Recently, suspension bioreactors have also been used to generate mini-brain-like structure from hPSCs for disease modeling, showing the important role of bioreactor in stem cell culture. This chapter describes a detailed culture protocol for neural commitment of hPSCs into neural progenitor cell (NPC) spheres using a spinner bioreactor. The basic steps to prepare hPSCs for bioreactor inoculation are illustrated from cell thawing to cell propagation. The method for generating NPCs from hPSCs in the spinner bioreactor along with the static control is then described. The protocol in this study can be applied to the generation of NPCs from hPSCs for further neural subtype specification, 3-D neural tissue development, or potential preclinical studies or clinical applications in neurological diseases.

Keywords: Pluripotent stem cell, Neural differentiation, Suspension culture, Bioreactor

---

1 Introduction

Human pluripotent stem cells (hPSCs), including human embryonic stem cells (hESCs) or human induced pluripotent stem cells (hiPSCs), hold great potential for establishing in vitro models for pathological disease study and drug discovery, and potentially generating therapeutic cells for cell therapy [1]. The conventional two-dimensional (2-D) culture on an adherent surface, however, cannot provide the large numbers of hPSCs and their derivatives required by the commercial and clinical applications [2]. For example, it is estimated that about 10^9–10^{10} hPSCs are needed to generate adequate quantities of tyrosine hydroxylase-positive neurons for Parkinson's disease treatment [3]. To provide the required amount of hPSCs for differentiation, three-dimensional (3-D) suspension culture platforms using bioreactors have been explored in recent years, based on the formation of cell aggregates or cultures with microcarriers [4, 5]. Although suspension culture of hPSCs using microcarriers have been successfully demonstrated, the additional separation of cell products from microcarriers at harvesting step would be cumbersome [6].

HPSCs are amendable for aggregate-based culture, as undifferentiated aggregates or embryoid bodies (EBs) for lineage-specific differentiation [7, 8]. Expansion of undifferentiated hPSC aggregates in suspension bioreactors has been well studied with different culture medium, split protocols, agitation speed, etc. [7, 9]. The use of suspension bioreactors for high-efficiency differentiation of specific lineages from hPSCs, however, needs to be further explored. For example, differentiation of hPSCs into oligodendrocyte progenitors requires EB culture for up to 4 weeks [10]. It is tedious to generate oligodendrocyte progenitors in low-attachment CellSTACKs. Adaptation of monolayer-based differentiation protocol from hPSCs into suspension culture has been investigated to allow for scalable production of cardiomyocytes [11], indicating the importance to generate hPSC-derived cells in a suspension bioreactor. For neural lineages, suspension bioreactors have been used for expansion of adult neural progenitor cell (NPC) spheres and agitation is able to control the aggregate size and minimize the appearance of necrosis center in the aggregates [12–14]. Spatial and temporal control of cell aggregation of hPSCs has also been observed to influence neural lineage commitment [15, 16].

Recently, a suspension bioreactor was used to generate brain organoids from hPSCs, the aggregate-like structure with definable forebrain, midbrain, and hindbrain layers [17]. The purpose of the bioreactor is to enhance the diffusion of nutrients and growth factors and allow for the generation of large size of neural spheres or mini-organs (up to 4 mm) for disease modeling [18]. The 3-D suspension culture of NPC spheres in bioreactors would benefit the generation of human cortical spheroids [19], while the influence of dynamic culture on neural tissue patterning in these 3-D cultures remains unclear. All together, the bioreactors are not simply a tool for culture scaleup, but also provide culture environment to construct in vitro 3-D tissue models.

In this chapter, a detailed procedure is described for the generation of hiPSC-derived 3-D NPC spheres in a suspension bioreactor. The neural induction from hiPSCs is induced by dual SMAD inhibition [20], using SB431542, a potent and specific transforming growth factor-β inhibitor, and LDN193189, a potent inhibitor of the bone morphogenetic protein pathway, based on the EB formation. Briefly, undifferentiated single hiPSCs are inoculated into a spinner bioreactor and maintained in neural differentiation medium in the presence of Rho kinase (ROCK) inhibitor Y27632 at the first day. After the EBs are formed, the cells are induced for neural differentiation for 7–21 days. The generated NPC spheres can be replated onto 1 % Geltrex-coated plates for further differentiation into specific neural cell types. The protocol in this study can be applied to the generation of NPCs from hPSCs for further neural subtype specification, 3-D neural tissue development, or potential preclinical or clinical applications in neurological diseases.

2 Materials

2.1 Materials for hiPSC Expansion to Seed Bioreactor Culture

1. *One vial of frozen undifferentiated hiPSCs.* The cell line used in this study is human iPSK3 cells (kindly provided by Dr. Stephen Duncan, Medical College of Wisconsin) [21, 22]. This cell line was derived from human foreskin fibroblasts transfected with plasmid DNA encoding reprogramming factors OCT4, NANOG, SOX2, and LIN28. The frozen vials are stored in liquid nitrogen tank for long-term preservation.

2. *1 % Geltrex-coated surface:* Geltrex LDEV-Free Reduced Growth Factor Basement Membrane Matrix (Life Technologies, #A1413202) is diluted at 1:1 with cold Dulbecco's Modified Eagle's medium (DMEM, Life Technologies, Carlsbad, CA). The aliquots are stored at −20 °C. Prior to coating, 1:1 diluted Geltrex is thawed at 2–8 °C overnight. Additional 1:50 dilution is made with cold DMEM with complete mixing before coating. The culture surface is coated with 1 % Geltrex at 0.1 mL/cm^2 at 37 °C for at least 1 h before the use (**Note 1**).

3. *Preparation of undifferentiated hiPSC culture medium:* mTeSR™1 (StemCell™ Technologies Inc., #05850) medium is used for undifferentiated hiPSC maintenance. This medium is composed of mTeSR™1 basal medium and 20 % serum-free supplement. The serum-free supplement aliquots are stored in −20 °C freezer for long-term use. The complete medium can be stored at 2–8 °C for 2 weeks (**Note 2**).

4. *Preparation of Accutase for cell dissociation:* Accutase (Life Technologies, #A1110501) is used to dissociate and passage hiPSCs grown on the Geltrex-coated surface, and it can dissociate hiPSK3 cells into single cells. Accutase aliquots are stored at −20 °C. After thawing, they are good for 2 weeks at 2–8 °C.

5. *Preparation of ROCK inhibitor Y27632:* ROCK inhibitor Y27632 (Sigma, #Y0503 or Stem Cell Technologies, #72302) is used to protect hiPSCs from apoptosis after dissociation into single cells. For stock solution, 1 mg Y27632 is dissolved in 294 μL of deionized (DI) water and stored at −20 °C for no longer than 6 months. The final working concentration is 10 μM by adding 1 μL of the stock solution per mL of culture medium.

2.2 Materials for hiPSC Differentiation into NPCs in a Spinner Bioreactor

1. *Preparation of NPC differentiation medium*: The differentiation medium contains DMEM-F12 plus 2 % B27 serum-free supplement (Life Technologies, #17504-044), referred as Differentiation Medium.

2. *Preparation of SMAD inhibitors for NPC differentiation*: Neural lineage differentiation is triggered by the inhibition of SMAD signaling pathway using two small molecules, SB431542 (Sigma, #S4317) and LDN193189 (Sigma, #SML0559). Both inhibitors are dissolved in DMSO (Sigma, #D2650) and kept at −20 °C for up to 3 months. For neural induction, SB431542 and LDN193189 stock solutions are added into the Differentiation Medium which results in working concentrations of 10 μM SB 431542 and 100 nM LDN 193189.

3. *Preparation of spinner bioreactors for NPC differentiation from hiPSCs:* A 50 mL glass spinner bioreactor (Wheaton, #356875) is used as an example to show NPC differentiation from hiPSCs in a dynamic suspension culture. Prior to the inoculation, the vessel is immersed with 2 mL Sigmacote (Sigma, #SL2) to avoid cell attachment on the glass surface. Then the vessel is dried at room temperature. The spinner bioreactor is autoclaved at 121 °C for 20 min before using (**Note 3**). Corning® disposable spinner flasks (Corning, #3152) can also be used for NPC differentiation from hPSCs depending on the experimental purpose.

4. *Wheaton Micro-Stir platform*: Wheaton Micro-Stir platform (Wheaton, #W900701-A) is recommended for the use with the bioreactor. The Micro-Stir platform allows for the control of agitation speed and the pattern (e.g., intermittent agitation with different cycles). The Micro-Stir platform can be put into a 37 °C, 5 % CO_2 incubator for bioreactor culture.

3   Methods

3.1 Expansion of Undifferentiated hiPSCs

3.1.1 Thaw the Vial of hiPSC from a Cell Bank

1. Remove a frozen vial of human iPSK3 cells from the cell bank and immediately place the vial in a 37 °C water bath (**Note 4**).

2. Gently swirl the cryovial to quickly thaw the cells until just a small piece of ice is left in the container.

3. Spray the outside of the vial with 70 % ethanol and open it in a biological safety cabinet.

4. Transfer the cell suspension into a 15 mL centrifuge tube, and dropwise add the thawing medium (mTeSR™1 basal medium) with ten times volume of cell suspension in the vial, and then centrifuge the cells at $300 \times g$ for 5 min.

5. After centrifugation, make sure that a complete cell pellet is visible at the bottom of the tube and gently aspirate the supernatant.

6. Resuspend the cells with 3 mL mTeSR™1 culture medium containing 10 µM ROCK inhibitor Y27632 and distribute the cells evenly into a Geltrex-coated 6-well plate at 1–1.5×10^5 cells/ cm². Place the plate in a 37 °C, 5 % CO_2 incubator.

3.1.2 Undifferentiated hiPSC Culture

1. Maintain hiPSCs on Geltrex-coated tissue culture 6-well plate in mTeSR™1 medium. Culture medium is replaced daily for 5–6 days. Check the culture confluency during the culture. For human iPSK3 line, the cells easily reach 100 % confluency at days 4–5 and become more packed at days 5–6.

2. For passaging, incubate human iPSK3 cells with 1 mL Accutase at 37 °C for 5 min (**Note 5**). Gently pipet the detached cell clumps for dissociation into single cells and transfer the single-cell suspension to a 15 mL tube containing 2 mL fresh medium.

3. Take a small amount of sample and count the cells using a hemocytometer. Based on the cell count, calculate the number of the harvested vessel(s) and the cell number that are required to seed the new vessels.

4. Spin down the cells at $300 \times g$ for 5 min and remove the supernatant. The cell pellet is resuspended in fresh culture medium and the cells are seeded at a density of 1.0–1.5×10^5 cells/cm² on Geltrex-coated surface in the presence of 10 µM Y27632 for the first 24 h. Place the culture in a 37 °C, 5 % CO_2 incubator.

5. The above procedure can be repeated until the targeted cell number is reached.

3.2 Differentiation of NPC Spheres from hiPSCs in a Suspension Bioreactor

3.2.1 Start the Suspension Culture with Single-Cell Inoculation on Day 0

1. After harvesting undifferentiated human iPSK3 cells, determine the cell number and volume to be seeded into the spinner bioreactor (**Note 6**). The seeding density is around 4.0–4.5×10^5 cells/mL. A working volume of 15 mL is used for a small-scale study. The static control is maintained in a low-attachment 24-well plate (Corning, #3473) seeded with the same cell density (i.e., 4.0–4.5×10^5 cells/mL).

2. Collect the required volume of hiPSCs in a sterile centrifuge tube. Centrifuge the tube at $300 \times g$ for 5 min and remove the supernatant.

3. Prepare 15 mL of Differentiation Medium containing 10 µM ROCK inhibitor Y27632 (**Note 7**). Resuspend the cell pellet with the prepared medium.

4. Transfer the cell suspension into the spinner bioreactor and place the bioreactor onto a magnetic stirrer (Micro-Stir, Wheaton, #W900701-A) in a 37 °C, 5 % CO_2 incubator (Fig. 1). Put the static control in the same incubator.

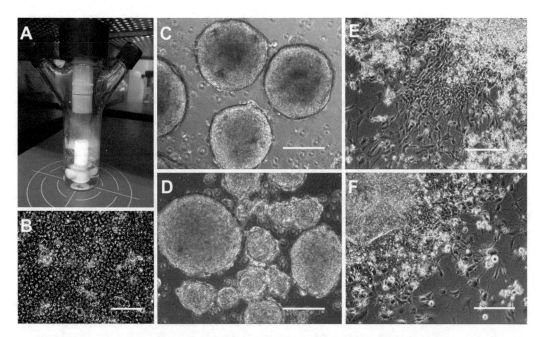

Fig. 1 Illustration of bioreactor culture of NPC spheres derived from hiPSCs. (**a**) Bioreactor setup. (**b**) Morphology of undifferentiated human iPSK3 cells, scale bar: 100 μm. (**c**) NPC spheres in suspension from bioreactor culture (**d**) NPC spheres in suspension from static culture (**e**) Replated NPC spheres from bioreactor culture. (**f**) Replated NPC spheres from static culture. Scale bar: 200 μm

5. Set up the agitation program of the spinner bioreactor. For better aggregate formation, the intermittent agitation is used right after the cell inoculation (**Note 8**). Initially, cells are agitated at a cycle with 15 min off and 15 min on at 80 rpm for ten cycles. After the intermittent agitation, continuous agitation at 80–100 rpm is used for the following culture (**Note 9**).

3.2.2 NPC Sphere Formation from hiPSCs in a Spinner Bioreactor During Days 1–7

1. Replace the medium on day 1. Briefly, collect the cell aggregates from spinner bioreactor into a 50 mL centrifuge tube. Let the aggregates set down at the bottom of the tube. Carefully remove the supernatant and resuspend the aggregates with fresh Differentiation Medium containing 10 μM SB431542 and 100 nM LDN193189. ROCK inhibitor Y27632 is removed at this point.

2. In-process monitoring of the bioreactor culture: The aggregates can be sampled every 2 days and imaged under a phase-contrast microscope to evaluate aggregate size distribution. The supernatant samples are taken for measuring glucose and lactate levels using YSI2700 Bioanalyzer to monitor the metabolic activities of the cells.

3. Change the medium every 2 days after day 1. Depending on the glucose and lactate levels, partial medium is changed during the culture period.

3.2.3 HiPSC-Derived NPC Sphere Culture in a Spinner Bioreactor During Days 7–20

1. After 7 days, the NPC spheres can be harvested for evaluation of progenitor markers. Or the culture can be continued.

2. For the continued culture, 10 μM SB431542 and 100 nM LDN193189 are replaced by different growth factors depending on the neuronal subtypes to be generated (**Note 10**). For example, for motor neuron differentiation, purmorphamine, fibroblast growth factor (FGF)-2, and retinoic acid can be used [23]. For cortical glutamatergic neurons, cyclopamine and FGF-2 can be used [24].

3.2.4 Further Differentiation of hiPSC-Derived NPC Spheres After Day 20

1. After 7–20 days of culture in a spinner bioreactor, the NPC aggregates are harvested in a 50 mL centrifuge tube. Aspirate the supernatant after the aggregates settle down.

2. Prepare differentiation medium without growth factors or with maturation factors such as brain-derived neurotrophic factor (BDNF). Resuspend the aggregates in the medium. Distribute the aggregates evenly into a Geltrex-coated 12-well plates or 24-well plates. The aggregates attach to the surface and are cultured for another 8 days in a 37 °C, 5 % CO_2 incubator.

3. The culture medium is changed every 2 days. Neuronal networks are visible under the microscope after the replating.

4. The cells can be characterized by various assays such as immunocytochemistry (Fig. 2), confocal microscopy, flow cytometry, and reverse transcription polymerase chain reaction (RT-PCR).

4   Notes

1. *Uniform coating with Geltrex:* The quality of Geltrex coating would affect cell expansion performance and NPC replating efficiency. The diluted Geltrex should be well mixed in cold DMEM before coating.

2. *Undifferentiated hiPSC culture medium:* HPSCs are sensitive to the quality of culture medium. The medium needs to be properly stored. The suboptimal culture medium would result in the cell detachment.

3. *Preparation of bioreactors:* Make sure that the paddle and magnet are assembled appropriately in the bioreactor. The agitation of the magnetic stirrer can be checked by pouring one-third of DI water in the vessel each time before use.

4. *Thawing hiPSCs:* The cryopreservation agent is detrimental to hiPSCs. Avoid keeping the cells in cryopreservation solution for a long time. In addition, hiPSCs are very fragile immediately post-thaw. It is better to transfer the cells to the desired culture environment as soon as possible.

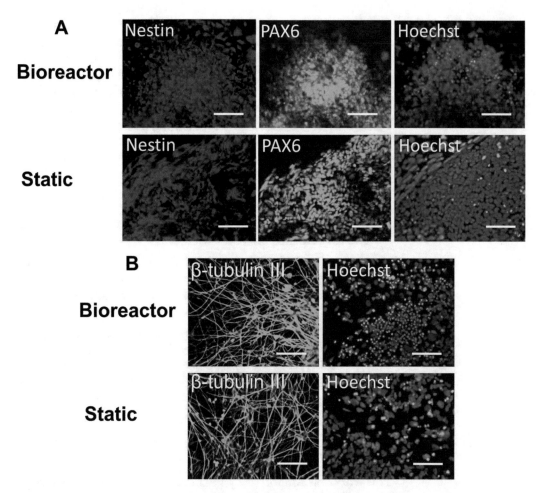

Fig. 2 Characterization of bioreactor-expanded NPC spheres derived from hiPSCs. The bioreactor-expanded NPC spheres were replated onto Geltrex-coated surface for 7 days. (**a**) Representative images showing neural progenitor markers Nestin and PAX6. (**b**) Representative images showing neuronal marker β-tubulin III. Scale bar: 100 μm

5. *Passaging hiPSCs using Accutase:* Closely monitor Accutase-induced hiPSC colony release from the culture surface. The duration of the treatment is crucial and over-incubation in the Accutase may damage the cells.

6. *Importance of starting cells for bioreactor culture:* The cell aggregation in the bioreactor is impacted by the quality and the composition of starting cells. The inoculated cells should easily form embryoid bodies in the suspension culture.

7. *ROCK inhibitor Y27632:* Y27632 is important for appropriate aggregate formation to form NPC spheres. In the absence of Y27632, human iPSK3 cells cannot form good EBs. It is only added to the culture medium after cell dissociation for the first 24 h. Prolonged exposure to Y27632 may affect cell proliferation.

8. *Agitation program:* The agitation strategy is very critical for good aggregate formation. To ensure homogeneous aggregate size distribution and to avoid large clumps, optimal agitation speed and intermittent agitation program should be used. It is suggested that the culture starts at a slightly low speed to avoid cell damage and to foster EB formation. Then the speed is slightly increased to keep the aggregates separated and avoid large cell clumps. After the aggregates are stabilized (e.g., after 4 days), the speed can be adjusted depending on the purpose of culture. For example, for NPC sphere expansion, agitation should be high enough to maintain NPC aggregate size less than 400 μm. For 3-D neural microtissue development, agitation speed can be reduced to minimize cell shedding and facilitate the size increase of microtissues.

9. *Agitation speed:* Agitation speed should be optimized depending on the spinner bioreactor geometry and the impeller shape. For different spinner bioreactors, there are different optimal ranges of agitation speed.

10. *NPC sphere commitment to specific neuronal subtypes:* Differentiation protocols can be modified based on the desired neuronal subtypes. In particular, different growth factor combinations should be optimized for the generation of different subtypes of neurons.

Acknowledgments

This work was supported by FSU start-up fund and partially from National Science Foundation (No.1342192).

References

1. Yu DX, Marchetto MC, Gage FH (2013) Therapeutic translation of iPSCs for treating neurological disease. Cell Stem Cell 12:678–688

2. Zweigerdt R (2009) Large scale production of stem cells and their derivatives. Adv Biochem Eng Biotechnol 114:201–235

3. Morizane A, Li JY, Brundin P (2008) From bench to bed: the potential of stem cells for the treatment of Parkinson's disease. Cell Tissue Res 331:323–336

4. Chen A, Reuveny S, Oh S (2013) Application of human mesenchymal and pluripotent stem cell microcarrier cultures in cellular therapy: achievements and future direction. Biotechnol Adv 31:1032–1046

5. Zweigerdt R, Olmer R, Singh H, Haverich A, Martin U (2011) Scalable expansion of human pluripotent stem cells in suspension culture. Nat Protoc 6:689–700

6. Wang Y, Cheng L, Gerecht S (2014) Efficient and scalable expansion of human pluripotent stem cells under clinically compliant settings: a view in 2013. Ann Biomed Eng 42:1357–1372

7. Serra M, Brito C, Correia C, Alves PM (2012) Process engineering of human pluripotent stem cells for clinical application. Trends Biotechnol 30:350–359

8. Kinney MA, Sargent CY, McDevitt TC (2011) The multiparametric effects of hydrodynamic environments on stem cell culture. Tissue Eng Part B Rev 17:249–262

9. Couture LA (2010) Scalable pluripotent stem cell culture. Nat Biotechnol 28:562–563

10. Li Y, Gautam A, Yang J, Qiu L, Melkoumian Z, Weber J et al (2013) Differentiation of oligodendrocyte progenitor cells from human embryonic stem cells on vitronectin-derived synthetic peptide acrylate surface. Stem Cells Dev 22:1497–1505

11. Kempf H, Olmer R, Kropp C, Ruckert M, Jara-Avaca M, Robles-Diaz D et al (2014) Controlling expansion and cardiomyogenic differentiation of human pluripotent stem cells in scalable suspension culture. Stem Cell Reports 3:1132–1146

12. Serra M, Brito C, Costa EM, Sousa MF, Alves PM (2009) Integrating human stem cell expansion and neuronal differentiation in bioreactors. BMC Biotechnol 9:82

13. Sen A, Kallos MS, Behie LA (2002) Expansion of mammalian neural stem cells in bioreactors: effect of power input and medium viscosity. Brain Res Dev Brain Res 134:103–113

14. Baghbaderani BA, Mukhida K, Sen A, Kallos MS, Hong M, Mendez I et al (2010) Bioreactor expansion of human neural precursor cells in serum-free media retains neurogenic potential. Biotechnol Bioeng 105:823–833

15. Miranda CC, Fernandes TG, Pascoal JF, Haupt S, Brustle O, Cabral JM et al (2015) Spatial and temporal control of cell aggregation efficiently directs human pluripotent stem cells towards neural commitment. Biotechnol J (in press). doi: 10.1002/biot.201400846

16. Steiner D, Khaner H, Cohen M, Even-Ram S, Gil Y, Itsykson P et al (2010) Derivation, propagation and controlled differentiation of human embryonic stem cells in suspension. Nat Biotechnol 28:361–364

17. Lancaster MA, Renner M, Martin CA, Wenzel D, Bicknell LS, Hurles ME et al (2013) Cerebral organoids model human brain development and microcephaly. Nature 501:373–379

18. Li Y, Xu C, Ma T (2014) In vitro organogenesis from pluripotent stem cells. Organogenesis 10:159–163

19. Pasca AM, Sloan SA, Clarke LE, Tian Y, Makinson CD, Huber N et al (2015) Functional cortical neurons and astrocytes from human pluripotent stem cells in 3D culture. Nat Methods 12:671–678

20. Chambers SM, Fasano CA, Papapetrou EP, Tomishima M, Sadelain M, Studer L (2009) Highly efficient neural conversion of human ES and iPS cells by dual inhibition of SMAD signaling. Nat Biotechnol 27:275–280

21. Si-Tayeb K, Noto FK, Sepac A, Sedlic F, Bosnjak ZJ, Lough JW et al (2010) Generation of human induced pluripotent stem cells by simple transient transfection of plasmid DNA encoding reprogramming factors. BMC Dev Biol 10:81

22. Si-Tayeb K, Noto FK, Nagaoka M, Li J, Battle MA, Duris C et al (2010) Highly efficient generation of human hepatocyte-like cells from induced pluripotent stem cells. Hepatology 51:297–305

23. Sun Y, Yong KM, Villa-Diaz LG, Zhang X, Chen W, Philson R et al (2014) Hippo/YAP-mediated rigidity-dependent motor neuron differentiation of human pluripotent stem cells. Nat Mater 13:599–604

24. Vazin T, Ball KA, Lu H, Park H, Ataeijannati Y, Head-Gordon T et al (2014) Efficient derivation of cortical glutamatergic neurons from human pluripotent stem cells: a model system to study neurotoxicity in Alzheimer's disease. Neurobiol Dis 62:62–72

Methods in Molecular Biology (2016) 1502: 129–142
DOI 10.1007/7651_2016_333
© Springer Science+Business Media New York 2016
Published online: 01 April 2016

Perfusion Stirred-Tank Bioreactors for 3D Differentiation of Human Neural Stem Cells

Daniel Simão, Francisca Arez, Ana P. Terasso, Catarina Pinto, Marcos F.Q. Sousa, Catarina Brito, and Paula M. Alves

Abstract

Therapeutic breakthroughs in neurological disorders have been hampered by the lack of accurate central nervous system (CNS) models. The development of these models allows the study of the disease onset/progression mechanisms and the preclinical evaluation of new therapeutics. This has traditionally relied on genetically engineered animal models that often diverge considerably from the human phenotype (developmental, anatomic, and physiological) and 2D in vitro cell models, which fail to recapitulate the characteristics of the target tissue (cell–cell and cell–matrix interactions, cell polarity, etc.). Recapitulation of CNS phenotypic and functional features in vitro requires the implementation of advanced culture strategies, such as 3D culture systems, which enable to mimic the in vivo structural and molecular complexity. Models based on differentiation of human neural stem cells (hNSC) in 3D cultures have great potential as complementary tools in preclinical research, bridging the gap between human clinical studies and animal models. The development of robust and scalable processes for the 3D differentiation of hNSC can improve the accuracy of early stage development in preclinical research. In this context, the use of software-controlled stirred-tank bioreactors (STB) provides an efficient technological platform for hNSC aggregation and differentiation. This system enables to monitor and control important physicochemical parameters for hNSC culture, such as dissolved oxygen. Importantly, the adoption of a perfusion operation mode allows a stable flow of nutrients and differentiation/neurotrophic factors, while clearing the toxic by-products. This contributes to a setting closer to the physiological, by mimicking the in vivo microenvironment. In this chapter, we address the technical requirements and procedures for the implementation of 3D differentiation strategies of hNSC, by operating STB under perfusion mode for long-term cultures. This strategy is suitable for the generation of human 3D neural in vitro models, which can be used to feed high-throughput screening platforms, contributing to expand the available in vitro tools for drug screening and toxicological studies.

Keywords: Human neural stem cells, Neural differentiation, Three-dimensional cultures, Stirred-tank bioreactors, Perfusion

1 Introduction

Human cell-based central nervous system in vitro models enable the study of healthy and pathophysiological features of neural cells in a human genome-based system, contributing to increase the accuracy of preclinical research [1]. Although primary human neurons present limited availability, a diversity of human neural cell sources have been made available, including immortalized cell lines or human neural stem cells (hNSC) from embryonic or adult origin or derived from human induced pluripotent stem cells (hiPSC) [2]. hNSC have self-renewal capacity that translates into in vitro expansion, while retaining the potential to generate cells from the three neural lineages—neurons, astrocytes, and oligodendrocytes [3]. Being isolated from specific fetal brain regions, such as forebrain or midbrain, these cells retain important developmental programs from their tissue of origin [4]. This confers an increased commitment towards specific neuronal lineages, as the case of human midbrain-derived NSC (hmNSC) that have been widely described to efficiently generate functional dopaminergic neurons [4–7]. hiPSC can be generated from somatic cells from any individual at any time point, which enables the generation of pluripotent cells carrying the genetic background from healthy donors or patients with specific mutations [8]. hiPSC can be directed towards neural commitment, giving rise to multipotent NSC [9]. Therefore the use of hiPSC-derived NSC (hiPSC-NSC) enables the study of how specific sets of mutations are related with the onset and progression mechanisms of neurological disorders, which was limited in non-iPSC models [8].

Three dimensional (3D) culture systems are able to mimic tissue architecture and complexity, namely in terms of cell–cell and cell–extracellular matrix interactions (ECM) [10, 11]. The generation of these 3D structures can be promoted by different methods, including the use of artificial matrices or scaffolds (e.g., hydrogels, fibrous meshes, porous sponges) or taking advantage of the potential of NSC to aggregate into neurospheres [11, 12]. Cell aggregation can be promoted either in static culture systems, as hanging-drop platforms and ultralow adhesion surfaces, or agitation-based systems, including orbital shakers and stirred-tank bioreactors [10]. As demonstrated by our group and others, agitation-based culture systems enable to efficiently drive hNSC differentiation as neurospheres into mature and functional neurons, astrocytes and oligodendrocytes [7, 13–15]. These systems present also increased mass transfer coefficients over static cultures, which minimize the formation of necrotic centers due to shortage of oxygen and nutrients in the inner layers [16].

The use of software-controlled stirred-tank bioreactors (STB) for NSC neurosphere culture has been reported to improve cell survival, proliferation and differentiation efficiency, relatively to 2D culture systems [17–19]. These systems provide tight control and online monitoring of physicochemical culture parameters, namely pH, dissolved oxygen and temperature. Importantly, STB culture can be operated under perfusion mode, which enables a continuous feed of fresh medium while removing spent medium and retaining the cells in the culture vessel. This continuous flow sustains more stable culture conditions in terms of nutrient and small molecules availability. Also by clearing cell debris and metabolic by-products, improves cell viability/functionality and provides a more physiological environment [20, 21]. These features contribute to a highly controlled culture system, which greatly improves the robustness and reproducibility of differentiation processes.

This chapter presents the protocols for aggregation and differentiation of hNSC, from fetal origin and derived from hiPSC, in perfusion-operated STB cultures. Cell aggregation in STB is highly dependent on cell type and culture vessel hydrodynamics. There are several parameters that modulate the hydrodynamics, including the geometry of the vessel, the type and size of the impeller, and stirring rate. Optimization of these parameters is crucial to sustain neurosphere formation, while minimizing detrimental effects of shear stress and allowing efficient gas/nutrient diffusion within the neurospheres to avoid the formation of necrotic centers [16]. Accurate hydrodynamic characterization of the systems provides data for process scale-up, essential for the production of the large cell numbers required in industrial and clinical settings. STB design is compatible with non-destructive sampling, allowing for a noninvasive characterization of long-term cultures throughout culture time [22]. Importantly, these features enable the use of STB cultures as feeding system of cell-based assays in high-throughput platforms.

2   Materials

All reagent preparation and cell culture handling should be carried out in a biological safety II cabinet under sterile conditions.

2.1   General Equipment

1. Biological safety cabinet (biosafety level II).
2. Inverted bright-field and fluorescence microscope.
3. Multi-gas incubator with CO_2 and O_2 control and humidified atmosphere.

2.2   Cell Culture and Characterization Supplies

1. Phosphate-buffered saline (PBS): 137 mM NaCl, 2.7 mM KCl, 10 mM $Na_2HPO_4 \cdot 2H_2O$, 1.8 mM KH_2PO_4 in ultrapure water. Adjust to pH 7.4, filter-sterilize using a 0.22 μm hydrophilic polyethersulfone (PES) filter and store for a maximum of 6 months at 4 °C.
2. Poly-L-ornithine (PLO) (Sigma, USA): prepare a stock solution of 1 mg/mL in sterile ultrapure water. Filter-sterilize using a 0.22 μm syringe PES filter and store at 4 °C.
3. Laminin (Sigma, USA): Store 1 μg/mL stock solution as aliquots at −20 °C. Always thaw laminin stock solutions on ice to avoid gelification.

4. Y-27632 ROCK inhibitor (Millipore, USA): dissolve in sterile ultrapure water to a concentration of 5 mM and store as aliquots at -20 °C.

5. Trypan blue working solution (Termo Fisher Scientific, USA): dilute 10 % stock solution in PBS to 0.1 % (v/v). Filter using a 0.45 μm syringe PES filter. The solution may be stored for up to 6 months at room temperature (i.e., 15–25 °C).

6. Fluorescein diacetate (FDA; Termo Fisher Scientific, USA): Prepare a stock solution by dissolving FDA in acetone in order to obtain a final concentration of 5 mg/mL. Store as aliquots at -20 °C.

7. Propidium iodide (PI) 1 mg/mL solution (Sigma, USA). Store as aliquots at 4 °C.

2.3 Bioreactor and Related Equipment

The protocol presented in this chapter is described for the use of DASGIP® Bioblock system equipped with the MP8 Multi Pump Module. Still, this procedure should be adaptable for other commercially available stirred-tank bioreactors systems, including single-use systems (e.g., Eppendorf BioBLU®; Sartorius UniVessel® SU).

1. Bioreactor system. 200 mL working volume bioreactor vessels (flat bottom with rounded corners), control unit and MP8 multi pump module from DASGIP Bioblock system (Eppendorf, Germany).

2. Paddle type impeller (trapezoid with arms as preferable geometry) and appropriate magnetic stirrer bar.

3. pH and dissolved oxygen (DO) probes.

4. Equipment for bioreactor assembly. Silicone tubing, autoclavable 0.22 μm polytetrafluorethylene (PTFE) filters, Luer connections, clamps, tweezers, glass flasks

5. Equipment for perfusion. Balances for gravimetric control, silicone tubing, perfusion tubes (0.5 mm), glass flasks for culture medium inlet and outlet lines, and stainless steel filter/sparger (10–20 μm pore size) for cell retention.

6. Equipment for sampling. Mobile biosafety cabinet or alternative equipment (i.e., sterile sampling devices) to ensure maintenance of culture sterility during inoculation or sampling operations. Sterile syringes (10 and 50 mL). Alternatively, disposable sterile sampling valves or automated bioreactor sampling systems can be used for aseptic sampling operations.

7. Reagents for cleaning and silanization of bioreactor glass vessels. Dichlorodimethylsilane, toluene, and potassium hydroxide (KOH).

2.4 Cell Culture Medium

Cell culture medium for expansion and differentiation typically differ according to the cell source, being derived from brain tissue or from hiPSC lines. Most common medium composition is based on the supplementation of a basal medium, as DMEM/F12 or Neurobasal. These supplementation strategies can be designed for untargeted neural differentiation, typically including insulin, progesterone, and sodium selenium [23], or based on a more directed approach for specific neuronal lineages, as FGF-8 and recombinant sonic hedgehog (SHH) for dopaminergic differentiation [24]. Complete culture medium should be filter-sterilized using a 0.22 μm vacuum filtration systems or syringe filters and store at 4 °C.

2.5 Biological Material

In this protocol, a procedure for aggregation and differentiation of hNSC derived from fetal midbrain or hiPSC is described. STB inoculation should be performed at 0.4×10^6 cell/mL. The working volume DASGIP Bioblock system ranges between 100 and 200 mL, which requires a minimum of 40×10^6 cells, up to 80×10^6 viable cells for inoculum.

3 Methods

Preparation of the STB units should be performed in advance, including bioreactor vessel preparation and assembly that should be initiated at least 2–3 days before inoculation following the workflow presented in Fig. 1. During these process the correct calibration of pH and DO probes are critical for ensuring the success of the culture.

3.1 STB Assembly and Preparation

1. Prepare the STB vessels by cleaning and silanizing (*see* **Note 1**).

2. On control software, start a new workflow by defining the number of vessels to use, using a template with control strategy and selecting pH, DO and pump calibration in preparation procedures.

3. Proceed to pH calibration by using standard buffer solutions of pH 7.0 for offset calibration and 9.21 for slope calibration.

4. Proceed to pump calibration, using silicone tubing and pump tubes of equal diameter as the ones that will be applied in the perfusion tubing lines, two beakers (reservoir and receptacle) and two balances, according to manufacturer's instructions.

5. Start STB assembly by inserting all components as represented in Fig. 2 in the upper cap, including pH and DO probes, connection for inlet and outlet perfusion lines, connection for sampling tube.

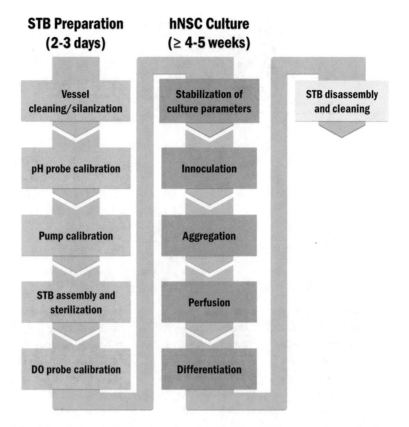

Fig. 1 General workflow for stirred-tank bioreactor (STB) preparation and culture of hNSC

6. For sampling connection, insert silicone tubing with a Luer connector in the extremity of the tube.

7. For perfusion, prepare one glass flask for inlet line and other for outlet line. For the inlet line add a silicone tube inside the glass flask to reach the bottom.

8. Connect glass flasks to STB vessel using appropriate silicone tubing and perfusion tubes.

9. If no dedicated system for exhaustion gas line cooling is available (e.g., off-gas condensers), a glass flask with water can be used for exhaustion gas bubbling. Connect it to the STB vessel using silicone tubing and one of the fixed STB upper cap connectors. A silicone tube should be also connected inside the flask for gas bubbling in water.

10. For aeration (headspace), connect silicone tube to other fixed STB cap connector, using a 0.22 μm filter in the extremity of the tube.

11. Assemble impeller parts.

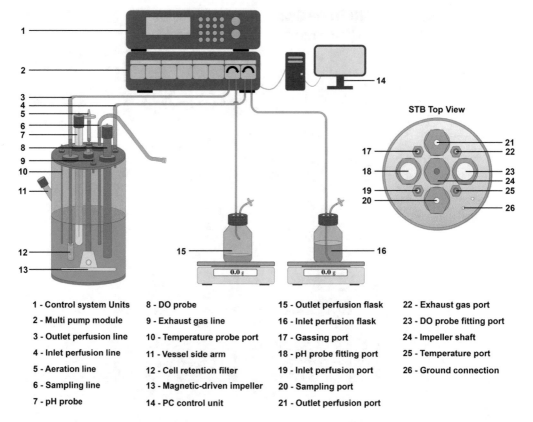

Fig. 2 Schematic representation of STB system used for hNSC culture and differentiation, operated under perfusion mode

12. Connect the cell retention inlet filter unit to the medium outlet port of the STB vessel.

13. Add ultrapure water to the vessel up to the level of the probes.

14. Check and tighten all fittings.

15. Apply clamps to all silicone tubing lines, except for the aeration line

16. Sterilize the STB and the connecting glass flasks by autoclaving for 30 min at 121 °C

17. After sterilization, let STB vessel to cool and check for damaged tubes or leakages.

18. Connect the STB to the control unit, by plugging in the DO and pH probes, inserting the temperature probe, connecting the aeration line and the grounding cable to the appropriate port in the STB cap.

19. In DO calibration interface, set temperature, agitation and aeration rates at working values (Table 1).

20. Allow DO probe to polarize (at least 6 h after connecting).

Table 1
Culture parameters setup for aggregation and differentiation of human neural stem cells

Working volume (mL)	Agitation rate (rpm)	pH	DO (%)	Temperature (°C)	Aeration rate (vvm)	Dilution rate (day^{-1})	Inoculum concentration (cell/mL)
100–200	70–100	7.4	15	37	0.1	0.33	4×10^5

21. Start DO calibration by defining the baseline oxygen level (i.e., zero-point calibration). Set the oxygen level setpoint to 0 %, wait for stable readings and calibrate offset. Define the air saturation level by setting the oxygen level setpoint to 21 % (i.e., 100 % air, with 21 % oxygen). Wait for stable readings and calibrate slope. Finish and leave DO calibration interface.

22. Remove and discard the ultrapure water through the sampling line. For the remaining volume use a 5 mL pipette to aspirate through the vessel side arm.

23. Add culture medium through the sampling port until 80 % of working volume.

24. Set culture parameters (Table 1) and switch on the unit controllers in the process view interface.

25. Proceed with cell inoculation once culture parameters (i.e., temperature, pH, and DO) are stabilized.

3.2 Bioreactor Inoculation

1. Expand hNSC as monolayers using standard culture protocols, which typically require a coated culture surface (e.g., PLO-laminin coating; *see* **Note 7**), presence of FGF2/EGF in the culture medium and low oxygen tensions (e.g., 3 % atmospheric oxygen) [4, 25, 26].

2. Detach confluent cell monolayers (e.g., using trypsin or other enzymatic methods), centrifuge and resuspend the cell pellet as single cell suspension in culture medium.

3. Determine cell concentration and viability by the trypan blue exclusion method (*see* **Note 5**). Determine the cell suspension volume required for inoculation of STB in order to attain the established inoculum (Table 1).

4. Dilute the cell suspension in fresh culture medium to a final volume equivalent to 20 % of the STB working volume (i.e., 40 mL if using 200 mL of working volume) and supplement with Y-27632 for a final concentration of 5 μM in the culture.

5. Proceed with the STB inoculation using the sampling line.

3.3 Initiating Perfusion Mode

Perfusion can be initiated at day 2 after inoculation, when the hNSC have aggregated into compact neurospheres with an average

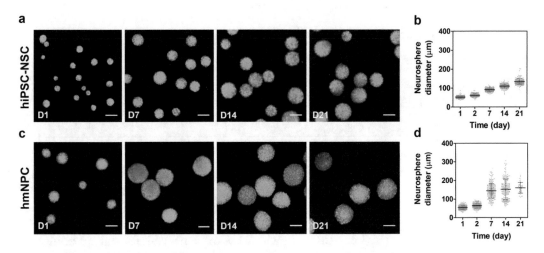

Fig. 3 hNSC viability and aggregation dynamics during 3D culture. (**a**, **c**) hiPSC-NSC (**a**) and hmNPC (**c**) neurosphere viability assay (fluorescein diacetate (FDA)—*green*; propidium iodide (PI)—*red*) along culture time. Scale bars, 100 μm. (**b**, **d**) Neurosphere diameter distribution of hiPSC-NSC (**c**) and hmNPC (**d**) as a function of time

diameter of 60–70 μm (Fig. 3). The retention of cells in perfusion systems typically rely on the use of size exclusion filters or cell density differences. In this protocol, a size exclusion strategy was followed by using a stainless steel sparger of 20 μm pore size ensuring the retention of the neurospheres in culture, while being able to wash out nonviable single cells and cell debris.

1. Place the perfusion tubes inside the respective pumps.

2. Fill the inlet flask with fresh aggregation medium.

3. Place inlet and outlet in the respective balances for gravimetric control.

4. Start pumps manually to fill up the inlet and outlet tubes. Be sure to use low flow rates to fill the outlet tube to prevent cell wash-out or filter clogging (typically 3–5 mL/h).

5. Determine the flow rate as a function of the dilution rate and working volume (*see* **Note 4**).

6. Start perfusion by setting the previously determined flow rates for the pumps.

7. Confirm the volumes added and withdrawn from the STB culture by using the gravimetric data. If required adjust the flow rate of the pumps.

8. Adjust the pump flow rates throughout culture time, according to the changes in the working volume due to sampling.

3.4 Bioreactor Disassembly

1. Stop process control interface.

2. Disconnect the probes and tubing from the control unit.

3. Sterilize the STB and the connecting glass flasks by autoclaving for 30 min at 121 °C

4. Clean and store the pH and DO probes according to the manufacturer's instructions.

5. Clean the STB vessel, impeller, and tubing with 70 % ethanol, ensuring that all medium residues have been removed to avoid growth of contaminants in the lines.

6. Rinse extensively all lines and STB parts with deionized water and let dry.

3.5 Culture Characterization

3.5.1 Cell Aggregation Monitoring

The aggregation period typically lasts 7 days until stabilization of neurosphere concentration is achieved, although this is a cell-line dependent process. The initial aggregation dynamics, within the first 1–8 h post-inoculation are critical, requiring frequent monitoring. Still, culture should be closely monitored up to day 2–3, ensuring that no large cell clumps are formed.

1. For monitoring cell aggregation, 0.5–1 mL samples are collected from the STB culture and observed in a bright-field microscope. A progression from single cells to duplets and triplets is observed within the first hours of culture. Typically after 8 h, the culture is mainly composed by multicellular neurospheres.

2. Use bright-field microscope images for determination of neurospheres diameter at each time point, as well as for counting the number of neurospheres to determine neurosphere concentration. This can be performed using open source software for image analysis, as ImageJ [27].

3. In case of heterogeneous neurospheres or fusion of neurospheres (forming large clumps), increase the agitation rate in a stepwise manner of 10 rpm at a time with sampling and observation at each increment (1–2 h of interval between steps; increase up to 100–110 rpm). High agitation rates should be avoided as these will increase shear stress, resulting in cell shedding at the neurospheres surface.

3.5.2 Cell Viability

1. Incubate neurospheres for 5 min with 20 µg/mL fluorescein diacetate, which stains viable cells, and 10 µg/mL propidium iodide, a membrane impermeable DNA dye that stains nonviable cells.

2. Image neurospheres using a fluorescence microscope equipped with the appropriate filters.

4   Notes

1. The silanization process is important to prevent cell adhesion to the glass vessel, by turning the glass surface less hydrophilic. Before silanization it is important to ensure that the vessel is

clean, which can be achieved through the action of KOH cleaning solution. This solution can be prepared by dissolving 56 g of KOH in 76 mL of distilled water, under constant agitation and in ice, as this is an exothermic reaction. After KOH is completely dissolved, fill to 1000 mL with 96 % ethanol and keep the solution at room temperature in an amber glass bottle for 2 weeks before use. This solution, as well as the silanization reagents, can be reutilized. In case precipitate formation in KOH cleaning solution after some utilization, this can be filtered using a paper filter. After cleaning the glass vessels with KOH cleaning solution during 16–24 h, the vessel should be washed thoroughly with running water. Once dried, STB glass vessels are silanized by adding a small volume of dichlorodimethylsilane and rotating the vessel ensuring that the reagent covers the entire vessel inner culture surface. Discard dichlorodimethylsilane and repeat the process with toluene. These two steps can be repeated to ensure the complete silanization of the culture surface. Once dried, wash vessels thoroughly through running water.

2. During pH and DO probes calibration it is important to ensure that the electrical current measured is within an acceptable range (according to manufacturer's instructions). If the measurements for the two-point calibration are not within this range, the probes should be replaced. The DO probe can be tested before sterilization in order to check if it meets the required standards in current measurement. For this, connect the probe to the control unit during DO calibration interface and wait for 5–10 min until it stabilizes. Check if the current is within the acceptable range. Afterwards, insert the probe into a small container (e.g., 50 mL centrifuge tube) along with a nitrogen gas source. Observe the decay in the DO and current values. Typically, after 10 min the value should be close to 1 % DO and 10 mV of current. In case the DO probe meets the required values (check manufacturer's instructions), proceed with sterilization.

3. Before sterilization of the assembled STB, cover all luer connections, filter ends, probe heads and temperature port with aluminum foil. Close all tubing with clamps to prevent filling tubing with water, except for the gas mixture inlet line that should remain opened to allow pressure release during the sterilization cycle.

4. The flow rate used for perfusion depends on the dilution rate and the current working volume. For a dilution rate of 0.33 day^{-1} and a working volume of 200 mL, the volume to be exchanged would be 2.75 mL/h (i.e., 0.33 day^{-1} × 200 mL = 66 mL/day = 2.75 mL/h). The determined flow rate should then be set as outlet/inlet pumps flow rate in the process control interface. The exchanged volumes should be confirmed using the gravimetric data.

5. The aeration rate in Table 1 is defined as gas volume flow per unit of liquid volume per minute (vvm). The gas flow volume used is therefore dependent on the working volume, being determined by the following equation:

$$\text{Aeration Rate (vvm)} = \frac{\text{Gas Flow Volume (mL/min)}}{\text{Working volume (mL)}}$$

6. For determination of cell concentration of single cell suspension to be used as inoculum, trypan blue exclusion method should be used. This allows also to discriminate between viable (bright cells) and nonviable cells (blue cells), as this is a membrane-impermeable dye that will only stain nonviable cells whose membranes are damaged. For cell counting, samples diluted in trypan blue working solution can be loaded into a hemocytometer (e.g., Fuchs-Rosenthal or equivalent) and visualized under a bright-field microscope.

7. To prepare PLOL-coated glass coverslips start by placing sterile microscope glass coverslips (13 mm) into the wells of a 24-well culture plate. Afterwards, prepare PLO working solution by diluting stock solution to 0.16 mg/mL in PBS (containing Ca^{2+} and Mg^{2+}). Add PLO working solution to the culture plate ensuring that the entire surface is covered and incubate for at least 3 h at 37 °C. Discard PLO solution and wash surfaces two times by adding PBS (containing Ca^{2+} and Mg^{2+}) for 10 min. Prepare freshly laminin working solution by diluting stock solution to 1 ng/mL in PBS (containing Ca^{2+} and Mg^{2+}). Remove PBS, add laminin working solution ensuring that the entire surface is covered and incubate for at least 3 h at 37 °C. Discard laminin solution and wash surfaces two times by adding PBS (containing Ca^{2+} and Mg^{2+}) for 10 min. For storage, PLOL-coated surfaces should be maintained with PBS (containing Ca^{2+} and Mg^{2+}) at 4 °C for 1 week maximum.

8. To handle 3D culture samples, only cut micropipette tips should be used in order to prevent mechanical damage to the neurospheres. For this, regular micropipette tips can be cut in 3–4 mm with a scissor, with posterior sterilization by autoclaving.

Acknowledgements

We gratefully acknowledge Dr Johannes Schwarz for the supply of hmNPC within the scope of the EU project BrainCAV (FP7-222992) and Dr Tomo Saric and Dr Eric J. Kremer for the supply of hiPSC-NSC lines. The authors also acknowledge João Clemente for the implementation and support on the perfusion operation mode in the bioreactors. This work was supported by: Brainvectors (FP7-286071), funded by the EU and PTDC/EBB-BIO/ 119243/2010, funded by Fundação para a Ciência e Tecnologia (FCT), Portugal. "iNOVA4Health—UID/Multi/04462/2013", a program financially supported by FCT / Ministério da Educação e Ciência, through national funds and cofunded by FEDER under the PT2020 Partnership Agreement is also acknowledged. DS, APT, and CP were recipients of a PhD fellowship from FCT, Portugal (SFRH/BD/78308/2011, PD/BD/52473/2014, and PD/BD/52202/2013, respectively).

References

1. Casarosa S, Zasso J, Conti L (2013) Systems for ex-vivo Isolation and culturing of neural stem cells. In: Bonfanti L. (ed) Neural stem cells – New perspectives. InTech. pp 3–28

2. Schüle B, Pera RA, Langston JW (2009) Can cellular models revolutionize drug discovery in Parkinson's disease? Biochim Biophys Acta 1792:1043–1051

3. Schwarz SC, Schwarz J (2010) Translation of stem cell therapy for neurological diseases. Transl Res 156:155–160

4. Storch A, Paul G, Csete M et al (2001) Long-term proliferation and dopaminergic differentiation of human mesencephalic neural precursor cells. Exp Neurol 170:317–325

5. Schaarschmidt G, Schewtschik S, Kraft R et al (2009) A new culturing strategy improves functional neuronal development of human neural progenitor cells. J Neurochem 109:238–247

6. Brito C, Simão D, Costa I et al (2012) 3D cultures of human neural progenitor cells: dopaminergic differentiation and genetic modification. Methods 56:452–460

7. Simão D, Pinto C, Piersanti S et al (2015) Modeling human neural functionality in vitro: three-dimensional culture for dopaminergic differentiation. Tissue Eng A 21:654–668

8. Zeng X, Hunsberger JG, Simenov A et al (2014) Concise review: modeling central nervous system diseases using induced pluripotent stem cells. Stem Cells Transl Med 3:1412–1428

9. Gage FH (2000) Mammalian neural stem cells. Science 287:1433–1438

10. Breslin S, O'Driscoll L (2013) Three-dimensional cell culture: the missing link in drug discovery. Drug Discov Today 18:240–249

11. Pampaloni F, Reynaud EG, Stelzer EHK (2007) The third dimension bridges the gap between cell culture and live tissue. Nat Rev Mol Cell Biol 8:839–845

12. Hopkins AM, DeSimone E, Chwalek K et al (2015) 3D in vitro modeling of the central nervous system. Prog Neurobiol 125:1–25

13. Terrasso AP, Pinto C, Serra M et al (2015) Novel scalable 3D cell based model for in vitro neurotoxicity testing: combining human differentiated neurospheres with gene expression and functional endpoints. J Biotechnol 205:82–92

14. Paşca AM, Sloan SA, Clarke LE et al (2015) Functional cortical neurons and astrocytes from human pluripotent stem cells in 3D culture. Nat Methods 12:671–678

15. Mariani J, Simonini MV, Palejev D et al (2012) Modeling human cortical development in vitro using induced pluripotent stem cells. Proc Natl Acad Sci U S A 109:12770–12775

16. Kinney MA, Sargent CY, McDevitt TC (2011) The multiparametric effects of hydrodynamic environments on stem cell culture. Tissue Eng Part B Rev 17:249–262

17. Serra M, Brito C, Costa EM et al (2009) Integrating human stem cell expansion and neuronal differentiation in bioreactors. BMC Biotechnol 9:82

18. Rodrigues CA, Fernandes TG, Diogo MM et al (2011) Stem cell cultivation in bioreactors. Biotechnol Adv 29:815–829

19. Baghbaderani BA, Mukhida K, Sen A et al (2010) Bioreactor expansion of human neural precursor cells in serum-free media retains neurogenic potential. Biotechnol Bioeng 105:823–833

20. Serra M, Brito C, Sousa MFQ et al (2010) Improving expansion of pluripotent human embryonic stem cells in perfused bioreactors through oxygen control. J Biotechnol 148:208–215

21. Li Z, Cui Z (2014) Three-dimensional perfused cell culture. Biotechnol Adv 32:243–254

22. Serra M, Brito C, Correia C et al (2012) Process engineering of human pluripotent stem cells for clinical application. Trends Biotechnol 30:350–359

23. Androutsellis-Theotokis A, Murase S, Boyd JD et al (2008) Generating neurons from stem cells. In: Weiner LP (ed) Neural stem cells. Humana, New York, pp 31–38

24. Kriks S, Shim J-W, Piao J et al (2011) Dopamine neurons derived from human ES cells efficiently engraft in animal models of Parkinson's disease. Nature 480:547–551

25. Reynolds BA, Weiss S (1996) Clonal and population analyses demonstrate that an EGF-responsive mammalian embryonic CNS precursor is a stem cell. Dev Biol 175:1–13

26. Reynolds BA, Tetzlaff W, Weiss S (1992) A multipotent EGF-responsive striatal embryonic progenitor cell produces neurons and astrocytes. J Neurosci 12:4565–4574

27. Schneider CA, Rasband WS, Eliceiri KW (2012) NIH Image to ImageJ: 25 years of image analysis. Nat Methods 9:671–675

Methods in Molecular Biology (2016) 1502: 143–158
DOI 10.1007/7651_2015_318
© Springer Science+Business Media New York 2015
Published online: 12 February 2016

Scalable Expansion of Human Pluripotent Stem Cell-Derived Neural Progenitors in Stirred Suspension Bioreactor Under Xeno-free Condition

Shiva Nemati, Saeed Abbasalizadeh, and Hossein Baharvand

Abstract

Recent advances in neural differentiation technology have paved the way to generate clinical grade neural progenitor populations from human pluripotent stem cells. These cells are an excellent source for the production of neural cell-based therapeutic products to treat incurable central nervous system disorders such as Parkinson's disease and spinal cord injuries. This progress can be complemented by the development of robust bioprocessing technologies for large scale expansion of clinical grade neural progenitors under GMP conditions for promising clinical use and drug discovery applications. Here, we describe a protocol for a robust, scalable expansion of human neural progenitor cells from pluripotent stem cells as 3D aggregates in a stirred suspension bioreactor. The use of this platform has resulted in easily expansion of neural progenitor cells for several passages with a fold increase of up to 4.2 over a period of 5 days compared to a maximum 1.5–2-fold increase in the adherent static culture over a 1 week period. In the bioreactor culture, these cells maintained self-renewal, karyotype stability, and cloning efficiency capabilities. This approach can be also used for human neural progenitor cells derived from other sources such as the human fetal brain.

Keywords: Human neural progenitor cells, Pluripotent stem cells, Scalable expansion, Xeno-free medium, Stirred suspension bioreactor

1 Introduction

Currently, millions of individuals worldwide suffer from the effects of neurodegenerative diseases such as Parkinson's disease, Multiple sclerosis, Huntington's disease, and spinal cord injuries which are characterized by the loss of neurons in the brain or spinal cord [1]. However, despite extensive research and significant advancement in medical and surgical care, neurological recovery is still largely limited. To date, significant advances in neural stem/progenitor cell therapy have created new hope for treatment of neurological disorders by employing different strategies such as the replacement of damaged cells, remyelination, neuronal restoration, bridging of lesion cavities, permissive environment for plasticity, releasing of neurotrophic factors and anti-inflammatory cytokines [2].

Over the last two decades, non-central nervous system tissue specific stem cells (e.g., bone-marrow derived mesenchymal stem cells, adipose tissue-derived stromal cells) [3], human pluripotent stem cells [4, 5], or mature somatic cells have been used to generated neural progenitors by the use of directed differentiation or transdifferentiation technology. These stem cells can be further induced to generate specialized neuronal fates [6, 7]. Among these sources, the development of efficient, defined protocols to generate human neural progenitor cells (hNPC) from pluripotent stem cells has gained increasing attention and witnessed considerable progress during last decade [8–10]. hPSCs can provide an available and reliable source for commercial production of specialized neurons with clinical potential in neurodegenerative diseases and facilitate patient-specific therapies [8–10]. Therefore, different groups have focused on developing robust, scalable culture systems for the large scale production of hNPC as 3D cell aggregates or on microcarriers in stirred suspension bioreactors under GMP conditions [11–14].

However, most of current protocols for scalable production of hNPCs and neural cells suffer from different technical issues such as undefined culture conditions, heterogeneous cell aggregate formation, and the necessity of using microcarriers or static cell aggregate formation systems such as AgrreWell™ to produce hNPCs that have a homogenous size distribution [9, 12, 14, 15]. Here, we describe a robust protocol for long-term maintenance and large scale expansion of hNPC as size-controlled aggregates under xeno-free culture conditions. This has been achieved by optimizing key important bioprocess parameters in hNPC expansion as 3D aggregates such as single cell inoculation density, aggregate formation and dissociation kinetics, and hydrodynamic conditions in a stirred suspension bioreactor. This protocol can provide a valuable platform for the design of robust bioprocessing technologies for producing clinically relevant numbers of hNPCs and their specialized subtypes in fully controlled stirred suspension bioreactors under xeno-free conditions, which will result in the realization of tremendous clinical and industrial potentials of these unique cells.

2   Materials

All materials (media, growth factors, proteins) are prepared by using specific instructions according to their data sheets. All media and reagents are prepared and maintained at 4 °C (unless indicated otherwise) for no greater than 1 week, as recommended. We use sterile Ca^{2+} and Mg^{2+}-free phosphate buffered saline (PBS⁻) buffer for all dilutions, unless otherwise identified in the text.

2.1 Chemicals

1. DMEM/F12 (Life Technologies, cat. no. 11330-032).

2. Dimethyl sulfoxide (DMSO, Sigma-Aldrich, cat. no. D2650).

3. Fetal bovine serum (FBS, Hyclone, cat. no. SH30071.03).

4. Insulin–transferrin–selenium (ITS, Life Technologies, cat. no. 41400-045).

5. Knockout serum replacement (KOSR, Life Technologies, cat. no. 10828-028).

6. L-glutamine (L-Gln, Life Technologies, cat. no. A2916801).

7. β-mercaptoethanol (Sigma-Aldrich, cat. no. M6250).

8. Nonessential amino acid solution (Life Technologies, cat. no. 11140-050).

9. Phosphate buffered saline without Ca^{2+} and Mg^{2+} (PBS$^-$, Life Technologies, cat. no. 21600-069).

10. Penicillin/streptomycin (Life Technologies, cat. no. 15070-063).

11. Y-27632 (ROCK inhibitor, Sigma-Aldrich, cat. no. Y0503).

12. Trypsin/EDTA (0.05 %/0.53 mM, Life Technologies, cat. no. 25300-054).

13. Human epidermal growth factor (EGF, Sigma-Aldrich, cat. no. E9644). We use in-house produced recombinant EGF (Royan Institute).

14. Human basic fibroblast growth factor (bFGF, Sigma-Aldrich, cat. no. F0291). We use in-house produced recombinant bFGF (Royan Institute).

15. CHIR99021 (Stemgent, cat. no. 04-0004).

16. Glucose (Sigma-Aldrich, cat. no. G0350500).

17. Sodium bicarbonate (Sigma-Aldrich, cat. no. S5761).

18. N2 (Life Technologies, cat. no. 17502-048).

19. B27 (Life Technologies, cat. no. 17504-044).

20. Sigmacote™ (Sigma-Aldrich, cat. no. SL2).

21. Agarose powder (Sigma-Aldrich, cat. no. A9045, cell culture tested).

22. GlutaMAX (Life Technologies, cat. no. 35050-061).

2.2 Disposables

1. Cell culture dishes, 60 mm Easy-Grip™ (Falcon, cat. no. T427-12).

2. Falcon™ 50 ml (Falcon, cat. no. 14-432-22) and 15 ml conical centrifuge tubes (Falcon, cat. no. 14-959-53A).

3. Syringes (5, 20, and 50 ml).

4. Pipettes (TPP 5, 10 ml, cat. no. 94005, 94010).

5. 0.22 μm pore size syringe filter (Orange, cat. no. 1520012).

6. Vacuum filtration set (TPP 250 ml, cat. no. 99250).

7. Pipette tips (0.5–10, 5–100, and 50–1000 ml).

8. 60 mm non-adhesive bacterial plates (Griner, cat. no. 628102).

9. Cryotubes (TPP 2 ml, cat. no. 89040).

10. 40 μm cell strainer (Corning, cat. no. CLS431750).

11. 96-well U-shaped bottom plate (TPP, cat. no. 92097)

2.3 Equipment

1. Inverted phase contrast microscope (4×, 10×, 20×, and 40× objectives, Olympus, CKX41).

2. Micropipettes (Eppendorf, 1–10, 10–100, and 100–1000 μl).

3. Pipettor (Eppendorf, Germany, order no. 3120000.909).

4. Laminar flow hood (Class I and II, Jal Tajhiz, Iran).

5. Hemocytometer (Neubaur, HBG, Germany).

6. Spinner Flask System (CELLSPIN, Integra Bioscience, cat. no. 183 001).

7. 100 ml Glass spinner flask with glass bulb impeller (Integra Bioscience, cat. no. 182 023).

2.4 Reagent Setup

1. *bFGF solution.* Prepare the bFGF solution in Tris-base buffer to achieve a 20 μg/ml final concentration (1000×). Divide the solution into 100 μl aliquots in cryotubes and freeze them at −70 °C. Aliquots can be stored for greater than 1 year without the loss of activity, however once thawed, they must be used within 1 week.

2. *EGF solution.* Prepare EGF solution using Tris-base buffer to achieve a 10 μg/ml final concentration (1000×). Divide the solution into 50 μl aliquots and freeze at −70 °C. Aliquots can be stored for greater than 1 year; however, they must be used within 1 week after thawing.

3. *CHIR99021 solution.* A 10 mM concentrated stock solution of CHIR99021 can be prepared by the addition of 430 μl DMSO to the entire vial's contents (2 mg lyophilized powder). Warm the solution to 37 °C and keep it for 2–5 min if sediment still present. For the cell culture, add the desired quantity of CHIR99021 working solution to the 37 °C medium to reach suitable final concentration of 3 μg/ml. Following reconstitution, store aliquots at −20 °C. Stock solutions are stable for 6 months when stored as directed (*see* **Note 1**).

4. *ROCK inhibitor (ROCKi), Y-27632.* In order to prepare a 10 mM ROCKi stock solution (1000×), add 5 mg of Y-27632 to 14.75 ml of cold double distilled water and sterilize the solution by using 0.22 μm pore size syringe filters. Divide the final solution into 100 μl aliquots in cryovials and store at −20 °C. After thawing, maintain the solution at 2–8 °C in the dark. Y-27632 is light sensitive and should be handled in subdued "yellow" lighting.

5. *hNPC medium in adherent culture.* In order to generate hNPC medium, we added 2.5 ml of KOSR, 0.5 ml L-Gln or Gluta-MAX, 1 % N2 supplement, 0.001 % B27, 0.5 ml penicillin/streptomycin, 0.5 ml nonessential amino acid, 20 ng/ml bFGF (*see* **Note 2**), and 20 ng/ml EGF (*see* **Note 2**) to 35.5 ml DMEM/F12 medium to make a final volume of 50 ml. The pH of the medium should be approximately 7.4. This medium should be stored at 4 °C. Prior to use, bFGF and EGF were added at the mentioned concentrations in order to reach the highest activity levels.

6. *hNPC medium in dynamic culture.* To 37.5 ml DMEM/F12 medium, add 0.5 ml L-gln or GlutaMAX, 1 % N2 supplement, 0.001 % B27, 1 % ITS, 0.5 ml penicillin/streptomycin, 0.5 ml nonessential amino acid, 20 ng/ml bFGF (*see* **Note 2**), 20 ng/ml EGF (*see* **Note 2**), 3 µM CHIR (*see* **Note 3**), and 6 % glucose 6 %. The medium pH should be approximately 7.4. The medium should be made up to a volume of 50 ml and stored at 4 °C. Prior to use, bFGF, EGF, and CHIR should be added at the abovementioned concentration in order to reach the highest activity levels.

7. *Freezing medium.* Mix 10 % DMSO and 50 % FBS or KOSR, and 40 % NSC expansion medium. The medium should always be prepared fresh on ice.

8. *Laminin preparation for coating.* Dilute laminin (1:1000) in PBS⁻ and place a sufficient amount of the working solution to coat the culture plates. This solution should remain in the culture plates for at least 1 h at 37 °C.

9. *Poly-L-ornithine preparation for coating.* Dilute poly-L-ornithine (1:5) in PBS⁻ and allow the necessary amount to remain in the culture plates for 1 h at 37 °C.

10. *Preparation of poly-L-ornithine and laminin-coated plates.* Place 2 ml of poly-L-ornithine working solution in 60 mm diameter culture plates and allow the solution to remain for 1 h at 37 °C. Remove this solution. Again, place 2 ml of laminin working solution in 60 mm diameter culture plates for 1 h at 37 °C. Remove the second solution and wash the plate with PBS⁻ before cell seeding.

3 Methods

3.1 Thawing and Expansion of hNPCs in Adherent Culture Conditions

1. Wash the poly-L-ornithine/laminin-coated plate with PBS⁻ before cell seeding.

2. Add 10 µM/ml of ROCKi Y-27632 to 1 ml of NPCs culture medium (medium for adherent culture) 1–2 h before cell thawing.

3. Remove a frozen hNPC vial from the nitrogen tank and maintain it at room temperature until a small piece of ice remain visible in the cryovial. Add 1 ml of expansion medium to the cryovial, then centrifuge it for 3 min at 2000 × g.

4. Gently aspirate the supernatant and add 1 ml of fresh NPCs expansion medium (medium for adherent culture) (*see* **Note 4**).

5. Pipette the medium two to three times until the medium becomes cloudy and the cells are homogenously dispersed in the medium.

6. Transfer the entire cell suspension from the cryovial onto 60 mm poly-L-ornithine/laminin-coated plates that contain 1 ml of hNPC expansion medium at a volume of 5×10^5 viable cells/ml.

7. Gently move the plate horizontally and vertically several times in order to distribute the cells evenly.

8. Incubate the cells under standard cell culture conditions [37 °C, 5 % CO_2, 95 % relative humidity (RH)] in a CO_2 incubator.

9. Renew the entire medium every other day until 80 % confluency (3–7 days).

3.2 Passage of hNPCs in the Adherent Culture Condition

1. Remove the medium from the culture plates.

2. Wash the hNPCs with 2 ml PBS⁻.

3. After 30 s, gently aspirate the supernatant and add 1 ml trypsin/EDTA and incubate the cells for 1 min.

4. Slowly remove trypsin/EDTA from the culture plates, then add 2 ml of hNPC expansion medium (adherent culture medium) and pipette the dissociated cell suspension one to three times in order to obtain dissociated single cells (*see* **Note 5**).

5. Transfer and equally divide the cell population onto two plates (*see* **Note 6**).

3.3 Transfer from the Static Suspension Culture to a Dynamic Suspension Culture System

3.3.1 Transfer from an Adherent Culture to a Static Suspension Culture System

1. Prepare a 60 mm diameter agarose-coated bacterial dish for the static suspension culture (*see* **Note 7**).

2. Add 10 μM of ROCKi (Y-27632) to the expansion medium of hNPCs that were cultured in adherent conditions a minimum of 1–2 h before detaching the cells from the plate.

3. Aspirate the hNPC medium and rinse the plate using PBS⁻.

4. Add 1 ml of pre-warmed trypsin/EDTA and incubate the cells for 1 min.

5. Gently remove the trypsin/EDTA (*see* **Note 5**). Check for cell dissociation under an inverted-phase contrast microscope.

6. Add 2 ml of expansion medium that contains 10 µM of ROCK inhibitor.

7. Detach hNPCs by pipetting.

8. Pipette the medium three to five times until the solution becomes cloudy and the cells disperse completely.

9. Check for cell dissociation under an inverted-phase contrast microscope.

10. Count the cells using a hemocytometer. Use the trypan blue exclusion method to assess cell viability.

11. Transfer the cells onto 60 mm non-adhesive bacterial plates that were prepared according to **step 1**. Plates should contain 5 ml of NPCs expansion medium in dynamic culture with 1×10^6 viable cells/ml inoculation density.

12. Incubate the cells under standard conditions (37 °C, 5 % CO_2, 95 % RH) as static suspension culture.

13. After 2 days of culture and formation of hNPC spheres, swirl the plates till all aggregates collect in the center of the plate. Gently remove 80–90 % of the old medium from the sides and add 5 ml of fresh NSCs expansion medium without the addition of ROCK inhibitor.

14. Renew the entire medium every other day for up to 5–7 days. The optimum day for passaging is dependent on the hNPC line and a sphere diameter that should be less than 250 µm (Fig. 1).

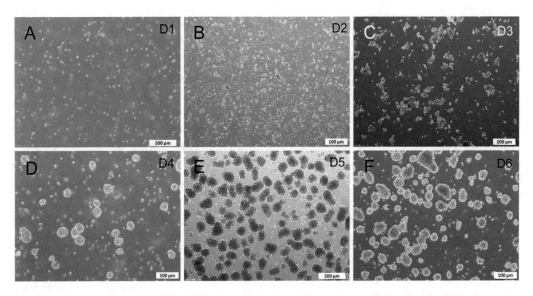

Fig. 1 Phase contrast microscopy of hNPC aggregate formation and growth kinetics in static suspension culture mode over a 6-day period

3.3.2 Procedure for the
Preparation of Glass
Spinner Flask Before Use or
at the End of Each Run

Before use or after the end of each run in a glass spinner flask, the spinner flask should be cleaned and siliconized in order to minimize cell attachment prior to inoculation with fresh NSCs.

1. Rinse the spinner flask multiple times with deionized water. If the flasks have been used, remove the spent medium that contains residual spheres and cells prior to washing with deionized water. The interior of the flasks should be thoroughly cleaned with a brush. Subsequently, rinse the flasks with 1 l of distilled water.

2. Fill the flasks with 70 % ethanol, maintain at in room temperature for at least 1 h, and subsequently rinse the flasks with 1 l deionized water after the ethanol is removed.

3. Fill the flasks with 1 N HCL and keep at room temperature for at least 30 min. Remove the HCl solution and rinse the flasks with 1 l deionized water.

4. Fill the flasks with 5 N NaOH and maintain at room temperature for a maximum of 12 h.

5. Remove NaOH and wash the flasks with at least 5 l water.

6. The flasks should be allowed to dry for 2–3 h at room temperature.

7. In order to minimize the risk of cell attachment to spinner flasks' internal surfaces during culture, the flasks should be siliconized by moistening the entire interior surfaces of the vessel that directly contacts the culture medium as well as the glass bulb impeller with 1–2 ml of Sigmacote™ under a fume hood. Excess Sigmacote™ solution can be removed after coating by using a 10 ml glass pipet or pipettor with a 1000 μl tip.

8. Dry the silicon coating by placing the spinners in an oven for 1 h at 120 °C or allowing them to remain overnight under a fume hood.

9. Rinse the spinner flasks three times with approximately 375 ml of deionized water at room temperature, and subsequently autoclave them after aspiration all of water from the vessel, putting in a thick nylon bag, and labelling with autoclave tape.

10. Moisten the interior of the flasks with 20 ml of hNSC medium before the hNPCs are added; exchange the medium with 50 ml of hNPC medium.

1. hNPC aggregates that have been cultured in the 60 mm non-adhesive bacterial plates according to above procedure (Sect. 3.3.1) should be passaged until they reach a suitable starting cell population (1×10^7 of hNPCs) to be inoculated into the spinner flask with 50 ml medium (2×10^5 viable cells/ml inoculation density).

2. Swirl the plates until the hNPC spheres collect in the center of the plate. Gently remove 80–90 % of the old medium from the sides and add 1 ml of trypsin/EDTA. Then, incubate the plates for 5 min.

3. Place the contents of the entire plate into a 15 ml test tube.

4. Centrifuge at 2000 × g for 3 min, and then gently remove the trypsin/EDTA.

5. Add 2 ml NPC expansion medium (hNPC medium in dynamic culture).

6. Detach the remaining hNPCs aggregates by pipetting.

7. Prepare spinner flasks (as described above). Seed NPCs at a density of 2×10^5 viable cells/ml (1×10^7 cells/50 ml) to 50 ml of complete hNPC expansion medium, after passing the single cell suspension through a 40 μm cell strainer to remove any possible aggregates that remained from the enzymatic dissociation (*see* **Note 8**).

8. Incubate the spinner flask under standard culture conditions (37 °C, 5 % CO_2, 95 % RH) by placing them on a magnetic stirring platform in a CO_2 incubator. Slightly loosen the side arm caps to allow for gas transfer. Set the agitation rate of the spinner flask to 45 rpm. By applying this agitation rate, a homogenous aggregate size (mean: 250 μm diameter) can be achieved which will largely facilitate passaging. The diameter and area of the hNPCs aggregates obtained from each culture mode should be measured under phase-contrast inverted microscope (Olympus) by using Olysia Bioreport software.

9. Renew the entire expansion medium (hNPC medium in dynamic culture) every other days, up to 5 days which is the optimum day for passaging, and according to the sphere diameters (Fig. 2) of the hNPC line (*see* **Note 9**).

3.3.4 Passaging hNPCs in the Dynamic Suspension Culture

1. Tighten the side arm caps, then transfer the spinner flask from the incubator to a laminar hood; wait for at least 15–20 min for sphere sedimentation. Collect all spheres in a 50 ml test tube (*see* **Note 10**).

2. Wash NPCs spheres once with PBS^-.

3. Remove the PBS^- and incubate the spheres for 5 min with a 2 ml trypsin/EDTA solution.

4. Centrifuge (3 min × 2000 × g), then gently remove trypsin/EDTA.

5. Add 2 ml expansion medium that contains 10 μM ROCK inhibitor.

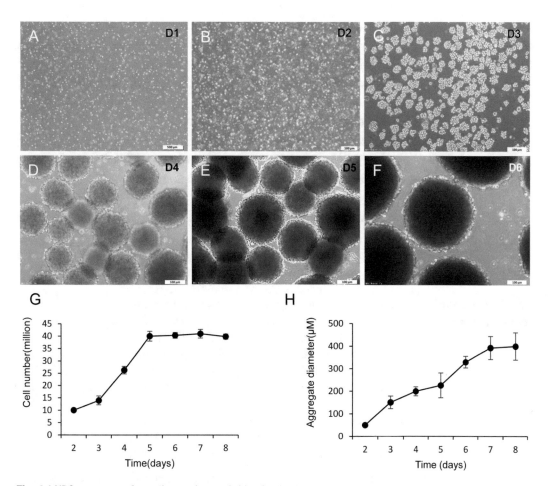

Fig. 2 hNPC aggregate formation and growth kinetics in the dynamic suspension bioreactor mode during 6 days. (**a–f**) Phase contrast microscopy. (**g**) Cell numbers. (**h**) Kinetics of the hNPC aggregate size

6. Pipette the NPCs clusters in suspension with a 2 ml pipette to reach a homogenous single cell suspension by observing under a microscope after pipetting.

7. Transfer the whole cell suspension onto a 60 mm bacterial dish and check cell dissociation under an inverted-phase contrast microscope.

8. Pass the single cell suspension through a 40 μm cell strainer to ensure no aggregates remain from the first step (*see* **Note 8**).

9. Count the cells using a hemocytometer. Assess cell viability by the trypan blue exclusion method.

10. Transfer 1×10^7 of NPCs to 50 ml of fresh NPC expansion medium (NPC medium in dynamic culture) to reach a 2×10^5 cell/ml seeding density. Incubate the spinner flask under standard culture conditions as previously described.

3.4   Freezing of hNPCs Generated in the Adherent and Dynamic Culture Systems

1. Passage the cells according to the appropriate passaging section (as explained in the adherent and dynamic culture systems) by using trypsin enzyme.

2. Add 1 ml of expansion medium for trypsin neutralization.

3. Count the cells using a hemocytometer. Assess cell viability by the trypan blue exclusion method.

4. Divide the correct amount of collecting medium into different cryovials to ensure the presence of 1×10^6 viable cells per vial.

5. Centrifuge vials at $1500 \times g$ for 5 min.

6. Gently aspirate the supernatant and add 900 µl of cold (4 °C) freezing medium (that contains 50 % KOSR and 40 % NPCs expansion medium).

7. Add 100 µl DMSO and transfer quickly to a −70 °C freezer for overnight storage. Next, transfer the cells to a nitrogen tank.

3.5   hNPC Colony-Forming Assay

The colony-forming assay is one of the most powerful tests to characterize neural stem cells and prove their capability to produce thousands of cells from one single cell. This method offers a number of advantages such as simplicity, reproducibility and the generation of an indefinite number of cells from a small number of cells in chemically defined serum free medium [16]. This assay is the best way to show hNPC self-renewal capability after culture in the bioreactor system. We have used this test to demonstrate that hNPCs maintained their self-renewal capability over multiple passages (ten passages) in the stirred suspension bioreactor.

1. Select an hNPC aggregate to be cultured in the spinner flask then trypisinize this aggregate (according to **step 4**) in a 15 ml test tube.

2. Prepare a minimum 50 wells of laminin/poly-L-ornithine-coated U bottom 96-well plates.

3. Dilute the single cell suspension with NPCs dynamic expansion medium until the minimum concentration of cells/ml of the medium is achieved.

4. Try to manually seed a single cell of NPCs per well of the U bottom coated 96-well plates.

5. Transfer the 96-well plates to a CO_2 incubator for culturing under standard conditions (37 °C, 5 % CO_2, 95 % RH).

6. Let cells to settle down and attach to the wells for 2–3 h.

7. Check the wells under the phase contrast microscope, then select single cell wells and mark them.

8. Renew the whole hNPCs dynamic expansion medium every other day, up to 20–25 days.

9. Mark the wells that have a single sphere (minimum: 50 μm diameter) as a successful colony forming assay for further characterization.

10. Conduct immunofluorescence staining to check for neural stem cell marker expressions of CD-133 and Nestin in select human hNPCs spheres (≥50 μm diameter size).

3.6 Anticipated Results

We evaluated the morphology of hNPCs and their quality attributes by analyzing the cells expanded in hNPC expansion medium under dynamic culture conditions over multiple passages (ten passages) for markers of hNPCs multipotency (Fig. 3a). hNPCs maintained their normal karyotype after ten passages in the dynamic suspension culture (Fig. 3b) and maintained their multipotency over long-term culture as determined by the expression of hNPC markers, *NESTIN*, *SOX1*, and *PAX6* (Fig. 3c). Additionally, after ten passages, these cells maintained their differentiation potential as detected by spontaneous differentiation into neuronal and glial cells (Fig. 4). We also assessed gene expression profiles during spontaneous differentiation in vitro and the colony forming capacity of hNPCs after expansion in the bioreactor system (Fig. 5). After freeze/thaw, hNPCs retained their aggregate formation properties and hNPCs marker expressions of NESTIN, PAX6 and SOX1 by immunofluorescence staining and kept their normal karyotype (data not shown).

We demonstrated that serial passaging of hNPC aggregates (ten passages) with this dynamic suspension protocol resulted in highly viable cell densities with good control of aggregate diameter size (250 ± 50 μm) which facilitated the passaging in dynamic culture condition and minimized necrosis in the center of the aggregates. The fold increase rate reached a maximum of 4.2 with 90 % viability over a period of 5 days compared to a maximum 1.5–2-fold increase in the adherent static culture over 1 week.

Therefore, the protocol presented here can offer a simple, robust and affordable platform for scalable expansion of hNPCs in suspension for a prolonged time, with high multipotency and karyotype stability.

4   Notes

1. For most cells, the maximum tolerance to DMSO is less than 0.5 %. CHIR is soluble in DMSO at a 100 mM concentration. We used a higher concentration (10 mM) of stock solution to reduce the effects of DMSO on live cells in the culture.

2. Dissolve the lyophilized bFGF/EGF in 1 ml of 20 mM Tris buffer (pH = 7). Importantly, growth factor working solution

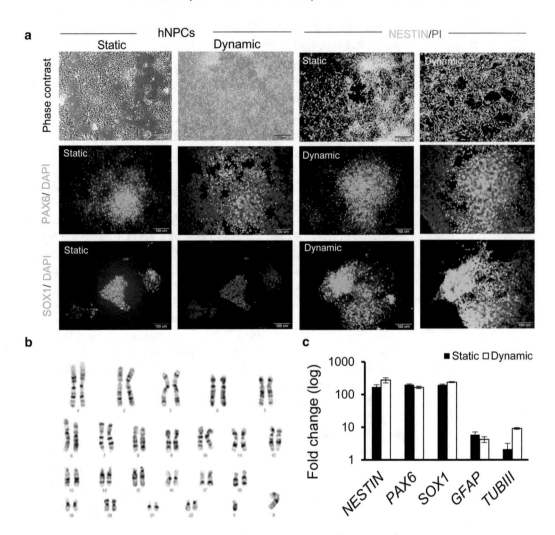

Fig. 3 Characterization of hNPCs before and after expansion in dynamic suspension culture in the stirred suspension bioreactor. (**a**) The morphology of replated hNPCs expanded in static suspension or dynamic suspension culture in the stirred suspension bioreactor observed with phase contrast microscopy. Expressions of neural progenitor cell markers (NESTIN, PAX6, SOX1) as detected by immunofluorescence staining. (**b**) Karyotype analysis shows chromosomal stability after expansion in the stirred suspension bioreactor. (**c**) Quantitative RT-PCR analysis of hNPC markers in static and dynamic culture systems

should not be kept for more than 1 week at 4 °C. We found that it best to prepare fresh growth factors each time.

3. Dissolve the lyophilized CHIR in DMSO or recommended solution according to the company datasheet and store at −20 °C.

4. The thawing process steps should be performed as quickly as possible, in a minimum amount of time (less than 3 min), because slower thawing causes large numbers of cells to die.

5. If the trypsin solution is cloudy, transfer it to a 15 ml tube and centrifuge for 3 min at 2000 × *g*, then remove the supernatant

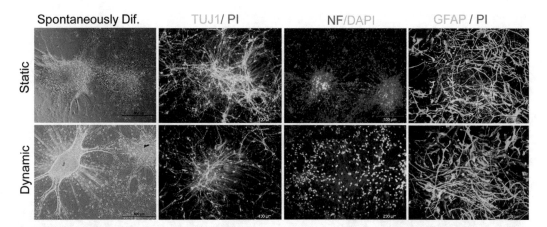

| Spontaneously Dif. | TUJ1/ PI | NF/DAPI | GFAP / PI |

Fig. 4 Characterization of spontaneously differentiated hNPCs in static and dynamic suspension cultures. Phase contrast image analysis and immunofluorescence staining for neuronal markers (TUJ1 and NF) and astrocyte marker (GFAP)

and add expansion medium. After pipetting, transfer the cells into culture plates.

6. Cell confluency is an important issue for hNPCs expansion. If the confluency is lower than 50 %, cells may begin to differentiate.

7. Dissolve 1 g of agarose powder (cell culture tested) in 100 ml PBS⁻ (1 %). Heat the solution to boiling and the powder completely dissolves. After cooling to 65–70 °C, coat the dishes with 2 ml of 1 % agarose and allow the plates to cool and gels solidify under a laminar hood.

8. If cell clumps are transferred to a spinner flask, they will cause heterogeneous, large aggregates to form, largely reducing the homogeneity of the culture system and cell quality since the diffusion rate of nutrients and metabolites inside large aggregates is largely limited [17, 18]. This phenomena can lead to increased necrosis, decreased viability, and inhibition of hNPCs proliferation.

9. Bring the spinner flask under the laminar hood and wait for 15–20 min (until all aggregates settle down). Then, incurve the spinner and slowly remove the above medium by a 10 ml sterile pipette. Continue to remove the medium until there is no risk of aggregate removal by medium aspiration. Finally, add 50 ml of fresh culture medium to the spinner flask.

10. Bring the spinner flask under the laminar hood and wait for 15–20 min (until all aggregates settle down). Then, incurve the spinner and attempt to collect all aggregates by a 10 ml sterile pipette or collect the entire spinner flask medium in a 50 ml tube.

a

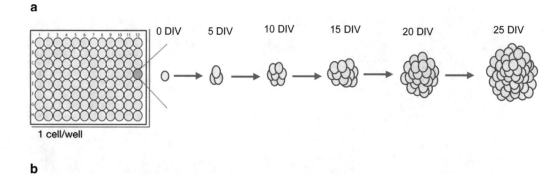

1 cell/well

0 DIV   5 DIV   10 DIV   15 DIV   20 DIV   25 DIV

b

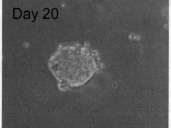

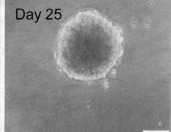

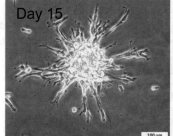

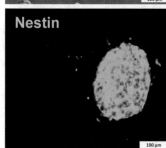

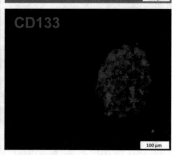

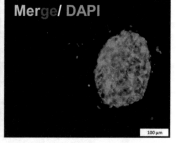

Fig. 5 Colony forming assay of hNPCs generated in the dynamic suspension culture. (**a**) Schematic illustration of the steps for the colony forming assay. (**b**) Phase contrast images of colony forming assay of 25 days culture in vitro and immunofluorescence staining for Nestin and CD133 after colony formation

Acknowledgments

This study was funded by grants provided from Royan Institute, the Iranian Council of Stem Cell Research and Technology, the Iran National Science Foundation (INSF), and Iran Science Elites Federation.

References

1. Chiu AY, Rao MS (2011) Cell-based therapy for neural disorders—anticipating challenges. Neurotherapeutics 8:744–752

2. Lunn JS, Sakowski SA, Hur J, Feldman EL (2011) Stem cell technology for neurodegenerative diseases. Ann Neurol 70:353–361

3. Ng TK, Fortino VR, Pelaez D, Cheung HS (2014) Progress of mesenchymal stem cell therapy for neural and retinal diseases. World J Stem Cells 6:111–119

4. Yan Y, Sart S, Li Y (2013) Differentiation of neural progenitor cells from pluripotent stem cells in artificial niches. Int J Stem Cell Res Transplant 1:22–27

5. Serra M, Brito C, Costa EM, Sousa MF, Alves PM (2009) Integrating human stem cell expansion and neuronal differentiation in bioreactors. BMC Biotechnol 9:82

6. Vierbuchen T, Ostermeier A, Pang ZP, Kokubu Y, Sudhof TC et al (2010) Direct conversion of fibroblasts to functional neurons by defined factors. Nature 463:1035–1041

7. Tian C, Ambroz RJ, Sun L, Wang Y, Ma K et al (2012) Direct conversion of dermal fibroblasts into neural progenitor cells by a novel cocktail of defined factors. Curr Mol Med 12:126–137

8. Nemati S, Hatami M, Kiani S, Hemmesi K, Gourabi H et al (2011) Long-term self-renewable feeder-free human induced pluripotent stem cell-derived neural progenitors. Stem Cells Dev 20:503–514

9. Baghbaderani BA, Mukhida K, Hong M, Mendez I, Behie LA (2011) A review of bioreactor protocols for human neural precursor cell expansion in preparation for clinical trials. Curr Stem Cell Res Ther 6:229–254

10. Pournasr B, Khaloughi K, Salekdeh GH, Totonchi M, Shahbazi E et al (2011) Concise review: alchemy of biology: generating desired cell types from abundant and accessible cells. Stem Cells 29:1933–1941

11. Hook L, Vives J, Fulton N, Leveridge M, Lingard S et al (2011) Non-immortalized human neural stem (NS) cells as a scalable platform for cellular assays. Neurochem Int 59:432–444

12. Miranda CC, Fernandes TG, Pascoal JF, Haupt S, Brüstle O et al (2015) Spatial and temporal control of cell aggregation efficiently directs human pluripotent stem cells towards neural commitment. Biotechnol J 10(10):1612–1624

13. Baghbaderani BA, Behie LA, Sen A, Mukhida K, Hong M et al (2008) Expansion of human neural precursor cells in large-scale bioreactors for the treatment of neurodegenerative disorders. Biotechnol Prog 24:859–870

14. Baghbaderani BA, Mukhida K, Sen A, Kallos MS, Hong M et al (2010) Bioreactor expansion of human neural precursor cells in serum-free media retains neurogenic potential. Biotechnol Bioeng 105:823–833

15. Bardy JA, Chen AK, Lim YM, Wu S, Wei S et al (2012) Microcarrier suspension cultures for high-density expansion and differentiation of human pluripotent stem cells to neural progenitor cells. Tissue Eng Part C Methods 19:166–180

16. Azari H, Rahman M, Sharififar S, Reynolds BA (2010) Isolation and expansion of the adult mouse neural stem cells using the neurosphere assay. J Vis Exp. doi:10.3791/2393

17. Abbasalizadeh S, Larijani MR, Samadian A, Baharvand H (2012) Bioprocess development for mass production of size-controlled human pluripotent stem cell aggregates in stirred suspension bioreactor. Tissue Eng Part C Methods 18:831–851

18. Mollamohammadi S, Taei A, Pakzad M, Totonchi M, Seifinejad A et al (2009) A simple and efficient cryopreservation method for feeder-free dissociated human induced pluripotent stem cells and human embryonic stem cells. Hum Reprod 24:2468–2476

Methods in Molecular Biology (2016) 1502: 159–168
DOI 10.1007/7651_2016_340
© Springer Science+Business Media New York 2016
Published online: 07 April 2016

A Microfluidic Bioreactor for Toxicity Testing of Stem Cell Derived 3D Cardiac Bodies

Jonas Christoffersson, Gunnar Bergström, Kristin Schwanke, Henning Kempf, Robert Zweigerdt, and Carl-Fredrik Mandenius

Abstract

Modeling tissues and organs using conventional 2D cell cultures is problematic as the cells rapidly lose their in vivo phenotype. In microfluidic bioreactors the cells reside in microstructures that are continuously perfused with cell culture medium to provide a dynamic environment mimicking the cells natural habitat. These micro scale bioreactors are sometimes referred to as organs-on-chips and are developed in order to improve and extend cell culture experiments. Here, we describe the two manufacturing techniques photolithography and soft lithography that are used in order to easily produce microfluidic bioreactors. The use of these bioreactors is exemplified by a toxicity assessment on 3D clustered human pluripotent stem cells (hPSC)-derived cardiomyocytes by beating frequency imaging.

Keywords: Microfluidics, Photolithography, Soft lithography, Human pluripotent stem cells (hPSCs), Organ-on-a-chip, Cardiomyocytes, Cardiac bodies (CBs), 3D cell culture models, Drug assessment

1 Introduction

The technology to manufacture bioreactors, with features at the scale of tens to hundreds of micrometers, originates from the production methods for microelectronics and microelectromechanical systems [1]. The ability of the technology to adapt to the scale and structure of biological systems and their applications is especially suitable for replicating the functions of living tissues and organs. Mimicking the in vivo environment is of major interest for establishing better prediction models during pharmaceutical development. With such models the assessment of efficacy and safety of drug candidates can be significantly improved, leading to better decisions for continuing drug development, than if conventional static 2D cell culture models are used [2–5]. Microfluidic bioreactors that consist of one or several cell types, and that are continuously perfused with cell culture media, providing nutrients and removing toxic wastes as well as inducing shear stress, are also known as organs-on-chips [6]. Important indicators of cellular state, such as essential proteins and metabolite markers, otherwise lost shortly after isolation, become easier to detect and monitor over extended time periods using these advanced cell culture platforms. Incorporating iPSCs with disease specific genotypes into organs-on-chips is expected to further improve the predictability of clinical trials [2, 7].

The most commonly applied fabrication method for microfluidic bioreactors is photolithography, followed by soft lithography. In photolithography, a master is created by spin coating a photoresist onto a silicon wafer that is polymerized by UV-light under a mask with transparent patterns. Soft lithography is then used for replica molding of the master in polydimethylsiloxane (PDMS), an optically transparent and gas permeable silicone elastomer. The photolithographic and soft lithographic fabrication methods are described in this chapter as well as the use of a microfluidic bioreactor with defined niches for drug assessment of beating clusters of hPSC-derived cardiomyocytes, known as cardiac bodies (CBs) [8–11].

2   Materials

Photolithography and soft lithography fabrication as well as cell culture require clean environments. Make sure that the laboratory is suitable for the production of the lithographic masters of the bioreactor and that all equipment used for culture of cells is sterile.

2.1   Photolithography

1. Silicon wafer, 4″ (SWI, Taiwan).

2. Washing solution (TL1): H_2O, H_2O_2 (VWR International, PA, USA) and NH_3 (Merck, NJ, USA) (5:1:1 ratio by volume).

3. Negative photo resists SU-8 10 and SU-8 3035 (MicroChem Corp., MA, USA).

4. Spin coater (Spin150, APT, Germany).

5. Photolithographic mask (Acreo, Sweden): The patterns of the bioreactor are created using any CAD software and printed on a plastic film. The desired structures should be transparent and the rest black.

6. Mask aligner with UV-light source (Kari Süss, Germany).

7. Developer (Micro Resist Technology, Germany).

8. Isopropanol (VWR International).

9. Silanization solution: 3 % tetramethylsilane (Sigma-Aldrich, MO, USA) in 99.5 % ethanol.

2.2   Soft Lithography

1. Silicon wafer patterned by photolithography (master).

2. Microscope glass slides (VWR International).

3. PDMS pre-polymer and curing agent (Sylgard 184, Dow Corning, MI, USA).

4. Punches (Syneo, TX, USA).

5. Bioreactor connections: stainless steel tubes (0.64 mm OD, New England Small Tube Corp., NH, USA), Tygon tubing (0.5 mm ID, 1.52 mm OD, VWR International), PEEK tubing (0.36 mm OD, LabSmith Inc., CA, USA).

2.3   Cell Culture

2.3.1   Production of Cardiac Bodies

1. Human pluripotent stem cells (hPSC) including the following human induced pluripotent stem cell (hiPSC) and human embryonic stem cell (hESC) lines: HES3 NKX2-5$^{eGFP/w}$ [12], hCBiPS2 [13], and hHSC_F1285T_iPS2 [14].

2. hPSC culture medium: mTeSR (Stem Cell Technologies, Vancouver, Canada) supplemented with 10 µM Y-27632 (Tocris, UK) for 24 h post seeding.

3. Cardiomyogenic differentiation medium RPMI1640 (Life Technologies, CA, USA) supplemented with B27 minus insulin (Gibco, USA) supplemented with the chemical WNT pathway agonist CHIR99021 (Millipore, MA, USA) and the WNT antagonist IWP2 or IWR1 (Tocris)

4. hPSC expansion and cardiomyogenic differentiation in suspension culture was performed [9, 11]. In brief: For the inoculation of suspension culture, dissociation into single cells is performed by accutase treatment for 5 min at 37 °C followed by dilution to 3.3×10^5 cells/ml in mTeSR1 plus 10 µM Y-27632 (Tocris) and seeding into 12-well suspension plates (Greiner-BioOne, Austria) or Erlenmeyer flasks (125 ml scale; VWR-International) in 1.5 ml/well or 20 ml/flask, respectively. Flasks are agitated at 75 rpm (orbital shaker, Infors-HT, Switzerland).

5. Differentiation of resulting aggregates is induced on day 0 (4 days after single cell inoculation) using 7.5 µM CHIR99021 (Millipore) for 24 h. On day 3, IWP2 or IWR1 (Tocris) is added at 5 and 4 µM for 48 h, respectively. During the differentiation, cells are kept in RPMI1640 (Life Technologies) supplemented with B27 (minus insulin). Medium is entirely replaced on days 0 (+CHIR 99021), 1, 3 (+IWP2 or IWR1), and 5 applying 3 or 20 mL depending on the respective culture format, i.e. 12-well dish and Erlenmeyer flask. Aggregates are cultured in RPMI1640 supplemented with B27 from day 7 onwards and used from day 10–30 for experiments.

6. Resulting cardiac bodies (CBs) are monitored by light microscopy; images are captured (AxiovertA1; Zeiss, Germany) and processed via the software AxioVision (Zeiss) to define diameter and size distribution ranging from ~200–400 µm.

1. Microfluidic bioreactor.

2. Cardiac bodies.

3. RPMI1640 supplemented with B27 minus insulin.

4. Petri dish (90 mm, VWR International).

5. Video microscope (SVM340, LabSmith Inc., CA, USA).

6. Syringes pump (Fusion 100, Chemyx Inc., TX, USA).

7. Three-port valve (MV201, LabSmith).

8. 1 ml syringe (BD Plastipak, NJ, USA).

9. 20 ml syringe (BD Plastipak).

10. Blunt ended cannulas (Sterican, B. Braun, Germany).

11. Crocodile clips.

12. Clamps.

3   Methods

3.1   Master Fabrication by Photolithography

Using the protocol presented here creates structures with thicknesses of 20 and 200 μm (Fig. 1). For other thicknesses check the specification of the photoresist and adjust spin times, baking times and UV-light exposure accordingly.

1. Immerse the silicon wafer in washing solution for 10 min at 85 °C to remove organic residues. Rinse the wafer in deionized water and let dry in an oven at 100 °C for 10 min.

2. Spin coat SU-8 10 on the wafer at 500 rpm for 10 s, then ramped at 150 rpm/s at 1500 rpm for 35 s to create the first layer of photoresist with a thickness of 20 μm (*see* **Notes 1** and **2**).

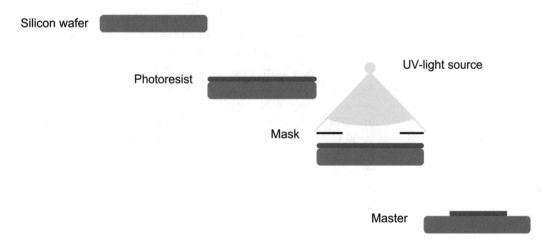

Fig. 1 Overview of the process of photolithography. A silicon wafer is covered with photoresist by spin coating. The photoresist is exposed to UV-light under a mask having a defined transparent pattern. UV-exposed photoresist is polymerized while unexposed photoresist is washed away

3. Remove solvents from the photoresist by baking the wafer at 65 °C for 1 min on a hot plate, followed by 10 min at 95 °C.

4. Expose the photoresist to UV-light at 187 mJ/cm^2 under the photolithographic mask.

5. Move the wafer to a hot plate at 100 °C for 15 min. The patterns should be visible after this polymerization step. Let the wafer cool down for 5 min before the next step.

6. Continue by spin coating SU-8 3035 on the wafer at 500 rpm for 10 s, then ramped at 150 rpm/s at 1500 rpm for 35 s to create a second layer of photoresist with a thickness of 100 μm. Bake the wafer at 65 °C for 1 min on a hot plate, followed by 100 °C for 65 °C.

7. Repeat **step 6** to create a third layer with a thickness of 100 μm (*see* **Note 1** and **2**).

8. Align the second photolithographic mask according to the patterns visible from the first exposure. Expose the photoresist to UV-light at 400 mJ/cm^2 (*see* **Note 3**).

9. Remove unexposed photoresist by immersing the wafer in developer for 10 min followed by rinsing in isopropanol, then immerse for another 2 min in developer, followed by rinsing in isopropanol and water.

10. Use a DekTak to check the height of the structures.

11. Put the wafer in a silanization solution for 10 min, rinse in 99.5 % ethanol and incubate at 120 °C for 30 min.

3.2 Microfluidic Bioreactor Fabrication by Soft Lithography

1. Mix PDMS pre-polymer and curing agent thoroughly at a 10:1 ratio (w/w) for at least 2 min. Degas with vacuum in a desiccator to remove bubbles (*see* **Note 4**).

2. PDMS membrane: Pour PDMS between two flat sheets of Plexiglas separated by a 200 μm diameter copper wire. Clamp the Plexiglas and let cure at room temperature overnight. Put in the oven at 60 °C for 10 min for final curing.

3. PDMS bubble trap: Pour 20 g of PDMS into a 90 mm petri dish and let cure overnight. Put in the oven at 60 °C for 10 min for final curing. Cut out a 5 × 5 × 4 mm cuboid and punch a hole through it. Cut a square of the plasma membrane and bond to the cuboid by exposing the surfaces to oxygen plasma for 1 min. Put the bubble trap in the oven for 10 min for irreversible bonding (*see* **Note 5**).

4. Clean the master in water and ethanol two times and blow dry. Pour the mixture of PDMS pre-polymer and curing agent over the master and cure in the oven at 60 °C for 2 h (Fig. 2).

5. Cut rectangles of PDMS from the master leaving at least 0.5 cm between the patterns and the edge.

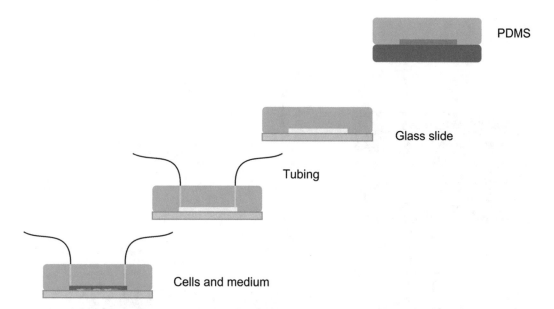

Fig. 2 Overview of the bioreactor production by soft lithography. PDMS is cured on the master, peeled off, and bonded to a microscope glass slide. Tubing is connected to the channel for cell seeding and cell culture medium perfusion by punching

6. Use tape to remove residues from the PDMS piece, rinse in water and ethanol and blow dry.

7. Use a 0.4 mm ID punch to make holes for the medium inlet.

8. Use a 0.64 mm ID punch to make holes for medium outlet, cell inlet and connection to the bubble trap.

9. Remove residues from the PDMS piece.

10. Oxidize the surfaces of the PDMS piece and a clean glass slide in oxygen plasma for 1 min. Bond the PDMS to the glass slide and make sure no air bubbles are trapped in between. Put in the oven at 60 °C for 10 min for irreversible bonding.

11. Remove residues from both the PDMS piece and the bubble trap membrane and oxidize in oxygen plasma for 1 min. Bond the bubble trap to the designated hole in the PDMS piece.

12. Attach the tubing to the medium inlet and outlet, the cell inlet and the bubble trap. Seal with PDMS and cure in the oven at 60 °C for 10 min (*see* **Note 6**).

13. Autoclave.

3.3 Beating Frequency Monitoring of 3D CBs in a Microfluidic Bioreactor

This step should be performed in a LAF hood. Place the device on top of a LabSmith inverted microscope and use it to detect bubbles, to monitor the CB seeding, and to record the beating frequency of CBs (Fig. 3).

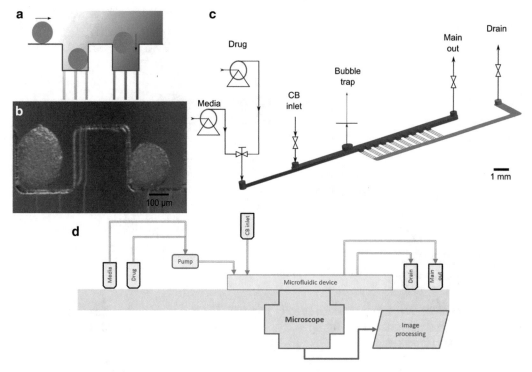

Fig. 3 (**a**) Schematics of the CB seeding process, (**b**) micrograph of CBs in bioreactor niches, (**c**) drawing of the bioreactor and connections, (**d**) the experimental setup as arranged in the incubator (reproduced with permission from ref. [8])

1. Maintain CBs in petri dishes with cell culture medium. Replace half of the medium every third day (*see* **Note 7**).

2. Insert blunt-ended cannulas into the Tygon tubing at the bubble trap.

3. Connect the PEEK tubing at the medium inlet to a LabSmith three-port valve.

4. Connect the other two ports of the valve to 1 ml syringes.

5. Punch a hole in two Eppendorf tubes and use as sample/waste at the end of the two outlet tubing.

6. Insert a 1 ml pipette tip in the Tygon tubing at the cell inlet.

7. Rinse the device with ethanol at 1 μl/min for 30 min. Purge any bubbles formed in the device by gently pushing the syringe (*see* **Note 8**).

8. Switch the inlet syringe to sterile water and rinse the device at 1 μl/min for 30 min.

9. Switch the inlet syringe to cell culture medium and rinse the device at 1 μl/min for 30 min.

10. Change one of the inlet syringes to drug supplemented medium.

11. Close the main medium outlet with a crocodile clip. Make sure that the medium only flows through the drainage outlet.

12. Sieve or manually pick 10 beating CBs with a diameter between 50 and 200 µm and seed them into the pipette tip at the cell inlet. Wait for a few minutes until the CBs reach the cell inlet. The CBs should follow the flow into separate niches (*see* **Note 9**).

13. Stop the pump and close the drain medium outlet and the CB inlet with crocodile clips. Cover the pipette tip with aluminum foil.

14. Move the whole setup to the incubator. Wait until the temperature has recovered before carefully opening the main medium outlet followed by the medium inlet. Start the syringe pump at a perfusion rate of 0.1 µl/min.

15. Record the beating frequency of the CBs.

16. To introduce the drug supplemented medium stop the pump. Gently switch the valve to the position of the drug supplemented medium syringe. Resume the perfusion with the new syringe.

17. Record again the beating frequency of the CBs.

4 Notes

1. Dispense approximately 1 ml of SU8 per inch of wafer diameter, i.e., 4 ml for a 4″ wafer. The first step of the spin coating is to distribute the photoresist over the wafer. Look at the wafer during this step to make sure that the photoresist is correctly dispersed.

2. The manufacturer's instructions for spin coating are made for photoresist on silicon. Spin coating a photoresist onto photoresist can give other thicknesses because of the friction change. If the thickness on your master is not approximately 200 µm revise this step and adjust the spin speed.

3. Alignment of the mask over the master can be difficult. A suggestion for increased precision is to make alignment marks of different sizes in the masks, for example crosses, spread over the mask.

4. The speed of PDMS curing depends on the temperature. In room temperature curing is completed after approximately 48 h. Store uncured PDMS in the freezer for reuse.

5. If many bioreactors should be made cut out a larger piece of the 4 mm thick PDMS, e.g., 5 × 80 × 4 mm in the petri dish. Punch holes approximately every 5 mm and bond a PDMS membrane to it. Cut out bubble traps when needed. This will save a lot of time.

6. Use a few days old PDMS for this step. It is stickier and will not trickle down and block the channels.

7. CBs have a tendency to stick together. Therefore, seed them sparsely and give the dish a swirl every day to keep properly sized material available.

8. The risk of bubble formation in the device can be reduced by equilibrating the bioreactor, the cell culture medium, the tubing and the syringes in the incubator over night before starting the experiment.

9. Be sure that the size of the CBs is smaller than the height of the bioreactor or they will be trapped. Gently tap the microscope if CBs stick together. Tilt the microscope if necessary to move the CBs.

Acknowledgements

The research leading to these results has received support from the Innovative Medicines Initiative Joint Undertaking under grant agreement n° 115439 (StemBANCC), resources of which are composed of financial contribution from the European Union's Seventh Framework Programme (FP7/2007-2013) and EFPIA companies' in kind contribution.

References

1. Whitesides GM (2006) The origins and the future of microfluidics. Nature 442 (7101):368–373. doi:10.1038/nature05058

2. Alépée N (2014) State-of-the-art of 3D cultures (organs-on-a-chip) in safety testing and pathophysiology. Altex. doi: 10.14573/altex1406111

3. Booth R, Kim H (2012) Characterization of a microfluidic in vitro model of the blood-brain barrier (muBBB). Lab Chip 12 (10):1784–1792. doi:10.1039/c2lc40094d

4. Toh YC, Lim TC, Tai D, Xiao G, van Noort D, Yu H (2009) A microfluidic 3D hepatocyte chip for drug toxicity testing. Lab Chip 9 (14):2026–2035. doi:10.1039/b900912d

5. Kaneko T, Nomura F, Hamada T, Abe Y, Takamori H, Sakakura T, Takasuna K, Sanbuissho A, Hyllner J, Sartipy P, Yasuda K (2014) On-chip in vitro cell-network pre-clinical cardiac toxicity using spatiotemporal human cardiomyocyte measurement on a chip. Sci Rep 4:4670. doi:10.1038/srep04670

6. Bhatia SN, Ingber DE (2014) Microfluidic organs-on-chips. Nat Biotechnol 32 (8):760–772. doi:10.1038/nbt.2989

7. Astashkina A, Mann B, Grainger DW (2012) A critical evaluation of in vitro cell culture models for high-throughput drug screening and toxicity. Pharmacol Ther 134(1):82–106. doi:10.1016/j.pharmthera.2012.01.001

8. Bergstrom G, Christoffersson J, Schwanke K, Zweigerdt R, Mandenius C-F (2015) Stem cell derived in vivo-like human cardiac bodies in a microfluidic device for toxicity testing by beating frequency imaging. Lab Chip 15 (15):3242–3249. doi:10.1039/C5LC00449G

9. Kempf H, Olmer R, Kropp C, Ruckert M, Jara-Avaca M, Robles-Diaz D, Franke A, Elliott DA, Wojciechowski D, Fischer M, Roa Lara A, Kensah G, Gruh I, Haverich A, Martin U, Zweigerdt R (2014) Controlling expansion and cardiomyogenic differentiation of human pluripotent stem cells in scalable suspension culture. Stem Cell Rep 3(6):1132–1146. doi:10.1016/j.stemcr.2014.09.017

10. Kensah G, Roa Lara A, Dahlmann J, Zweigerdt R, Schwanke K, Hegermann J, Skvorc D, Gawol A, Azizian A, Wagner S, Maier LS, Krause A, Drager G, Ochs M, Haverich A,

Gruh I, Martin U (2013) Murine and human pluripotent stem cell-derived cardiac bodies form contractile myocardial tissue in vitro. Eur Heart J 34(15):1134–1146. doi:10. 1093/eurheartj/ehs349

11. Kempf H, Kropp C, Olmer R, Martin U, Zweigerdt R (2015) Cardiac differentiation of human pluripotent stem cells in scalable suspension culture. Nat Protoc 10(9):1345–1361. doi:10. 1038/nprot.2015.089

12. Elliott DA, Braam SR, Koutsis K, Ng ES, Jenny R, Lagerqvist EL, Biben C, Hatzistavrou T, Hirst CE, Yu QC, Skelton RJ, Ward-van Oostwaard D, Lim SM, Khammy O, Li X, Hawes SM, Davis RP, Goulburn AL, Passier R, Prall OW, Haynes JM, Pouton CW, Kaye DM, Mummery CL, Elefanty AG, Stanley EG (2011) NKX2-5(eGFP/w) hESCs for isolation of human cardiac progenitors and cardiomyocytes. Nat Methods 8(12):1037–1040. doi:10. 1038/nmeth.1740

13. Haase A, Olmer R, Schwanke K, Wunderlich S, Merkert S, Hess C, Zweigerdt R, Gruh I, Meyer J, Wagner S, Maier LS, Han DW, Glage S, Miller K, Fischer P, Scholer HR, Martin U (2009) Generation of induced pluripotent stem cells from human cord blood. Cell Stem Cell 5(4):434–441. doi:10.1016/j.stem. 2009.08.021

14. Hartung S, Schwanke K, Haase A, David R, Franz WM, Martin U, Zweigerdt R (2013) Directing cardiomyogenic differentiation of human pluripotent stem cells by plasmid-based transient overexpression of cardiac transcription factors. Stem Cells Dev 22 (7):1112–1125. doi:10.1089/scd.2012.0351

Methods in Molecular Biology (2016) 1502: 169–179
DOI 10.1007/7651_2016_341
© Springer Science+Business Media New York 2016
Published online: 05 April 2016

Novel Bioreactor Platform for Scalable Cardiomyogenic Differentiation from Pluripotent Stem Cell-Derived Embryoid Bodies

Sasitorn Rungarunlert, Joao N. Ferreira, and Andras Dinnyes

Abstract

Generation of cardiomyocytes from pluripotent stem cells (PSCs) is a common and valuable approach to produce large amount of cells for various applications, including assays and models for drug development, cell-based therapies, and tissue engineering. All these applications would benefit from a reliable bioreactor-based methodology to consistently generate homogenous PSC-derived embryoid bodies (EBs) at a large scale, which can further undergo cardiomyogenic differentiation. The goal of this chapter is to describe a scalable method to consistently generate large amount of homogeneous and synchronized EBs from PSCs. This method utilizes a slow-turning lateral vessel bioreactor to direct the EB formation and their subsequent cardiomyogenic lineage differentiation.

Keywords: Bioreactor, Cardiomyocyte, Embryoid body, Pluripotent stem cells, Slow turning lateral vessel

1 Introduction

Approaches involving cardiomyocyte differentiation from pluripotent stem cells (PSCs) provide unique opportunities to produce sufficient cell numbers for cardiac regenerative therapy, tissue engineering, and drug screening for cytotoxicity testing [1, 2]. To establish these approaches for clinical and industrial applications, reliable and reproducible methodologies must be developed to ensure a scalable generation of differentiated PSCs for producing clinically relevant numbers of cardiomyocytes that are safe for human use [3].

Differentiation of PSCs into cardiomyocytes usually needs cell aggregation into spherical-like structures referred to as embryoid bodies (EBs), which are usually composed of three germ layers including the ones generating cardiac-like cells [4]. Although the hanging drop (HD) is able to produce homogeneous EBs (in terms of a regular shape and uniform size), this technique cannot produce EBs on a highly scalable manner, due to a labor-intensive procedure [5]. On the other hand, the static suspension culture (SSC) technique is easier to perform, but it shows poor reliability in the cell aggregation step, leading to large and irregularly shaped EBs causing cellular apoptosis in their center core [6]. Therefore, the latter two methodologies are currently not suitable for clinical and industrial applications.

The slow-turning lateral vessel (STLV) bioreactor has been developed to overcome the limitations with the previously mentioned culture systems [7]. The STLV bioreactor allowed for direct PSC aggregation under depleted fluid shear stress, high mass transfer, microgravity, and oxygen diffusion in testing studies [8]. Our laboratory has developed a protocol for scalable EB formation and cardiomyogenic differentiation from mouse PSCs by using this STLV bioreactor. The protocol requires an initial seeding density of 3×10^5 PSCs/mL at 10 rotations per minute (rpm) and plating the EBs onto gelatin-coated dishes on the 3rd day of culture. This results in EB production at a large scale and also an increased potential for differentiation into cardiomyocytes [9]. Moreover, the STLV bioreactor is able to produce homogeneous EBs with tunable sizes and shapes [10]. When compared with the SSC technique, the STLV bioreactor system significantly enhanced the efficiency of EB formation by increasing the yield of EB formation/ mL and the total cell yield of EB/mL by fourfold and sixfold, respectively; and also by reducing by 1 log the number of cells that were not incorporated into EBs [10]. This has been in agreement with another study [11]. However, this bioreactor system may produce variations in the size and shape of EBs in specific cell types such as human embryonic stem cells (ESCs) [12]. These heterogeneous EBs are generated perhaps due to (1) interruptions in the stirring process, (2) the microgravity within the STLV during media changes causing agglomeration of EBs, or (3) the cell seeding density and rotation speed not being fully optimized [9]. Additionally, STLV-derived EBs showed a larger beating area and a faster beating by enhancing up to 4.2 times the area of cardiac troponin T (cTnT) per EB when compared to SSC- and HD-derived EBs. These observations revealed that the STLV bioreactor methodology delivers scalability, reproducibility, and a relative simplicity in EB production and provides a striking efficiency in cardiomyogenic differentiation over the SSC and HD approaches [10].

2 Materials

2.1 Cell Lines

1. Mouse PSC lines (induced pluripotent stem cell (iPSC) or ESC lines) (see **Note 1**).

2. Feeder cells (Mitomycin C-inactivated confluent mouse embryonic fibroblast (MEF)).

2.2 Chemicals and Culture-Grade Reagents

1. High-glucose Dulbecco's modified Eagle medium (DMEM, Invitrogen, cat. no. 11965-092).

2. Fetal bovine serum (FBS, Hyclone, cat. no. SH30070.03) (*see* **Note 2**).

3. GlutaMAX™ Supplement (Invitrogen, cat. no. 35050-061).

4. β-mercaptoethanol (β-ME, Invitrogen, cat. no. 21985-023).

5. Nonessential amino acid solution (NEAA, Invitrogen, cat. no. 11140-050).

6. Penicillin–streptomycin (Invitrogen, cat. no. 15140-122).

7. Mouse leukemia inhibitory factor (mLIF, ESGRO, Chemicon International, cat. no. ESG1107).

8. L-ascorbic acid (Sigma, cat. no. A4403).

9. Trypsin (0.05 %) with ethylenediaminetetraacetic acid (EDTA) (Invitrogen, cat. no. 25300054).

10. Phosphate buffered saline without Ca^{2+}/Mg^{2+} (PBS, Invitrogen, cat. no. 10010-023).

11. Gelatin (Sigma, cat. no. G1890).

12. Trypan blue solution (0.4 % (wt/vol), Sigma Life Science, cat. no. T8154).

13. 70 % ethanol (ethyl alcohol).

14. Distilled water.

15. Labware detergent (e.g., 7× detergent).

2.3 Disposables

1. Cell culture dishes (60 and 100 mm, BD Falcon, cat. nos. 353002 and 353003).

2. Bacteriological petri dish (60 and 150 mm, BD Falcon, cat. nos. 351007 and 351058).

3. 24-well plate (Nunclon®, Sigma, cat. no. D7039).

4. Conical tubes (15 and 50 mL, Falcon, cat. nos. 352097 and 352098).

5. Serological pipettes (5, 10, and 25 mL, Thermo Scientific Nunc, cat. nos. 170355, 170356, and 170357).

6. Syringe (5 and 10 mL).

7. Rapid-Flow™ Sterile Disposable Filter Units (500 mL Bottle Top Filter with 75 mm Supor®machV PES Membrane, Thermo Scientific Nalgene®, cat. no. 595-4520).

2.4 Equipment

1. Slow-turning lateral vessel bioreactor (STLV, RCCS-4H with 110 mL of vessels, Synthecon, Cellon S.A. Bereldange, Luxembourg) (*see* **Note 3**).

2. Inverted phase-contrast microscope (Olympus, IX 71 equipped with an Olympus DP70 digital camera).

3. Stereomicroscope (Nikon, AZ100).

4. Laminar flow hood (Class II, Thermo Scientific).

5. Humidified incubator (37 °C, 5 % CO_2, Thermo Scientific).

6. Water bath (WNB 14, Memmert).

7. Centrifuge (Eppendorf, 5804R).

8. Hemocytometer, Neubauer improved (Carl Roth, cat. no. T728.1).

9. One-way stopcock with female Luer lock port and SPIN-LOCK® connector (B. Braun, cat. no. D100).

10. Laboratory glass bottle (100 and 500 mL, DURAN Group, cat. nos. 21801245, 21801445).

11. Micropipette (1–10, 10–100, and 100–1000 μL, Eppendorf, cat. nos. 3121000023, 3121000074, 3121000120).

12. Pipettor (Thermo Fisher Scientific, Matrix, CellMate II, cat. no. 9501).

2.5 Reagent Setup

1. Incomplete PSC medium: DMEM containing 15 % (vol/vol) FBS, 0.1 mM β-ME, 0.1 mM NEAA, 2 mM GlutaMAX™, and 1 % penicillin–streptomycin (50 μg/mL streptomycin and 50 U/mL penicillin). Filter-sterilize and store at 4 °C for up to 2 weeks (*see* **Note 4**).

2. Complete PSC medium: incomplete PSC medium plus 1000 or 2000 U/mL mLIF. Store at 4 °C for up to 2 weeks (*see* **Note 5**).

3. Cardiac differentiation medium: incomplete PSC medium supplemented with ascorbic acid at 50 μg/mL. Store at 4 °C for up to 2 weeks.

4. L-ascorbic acid: dissolve it in sterile water at 50 mg/mL (1000×). Then filter-sterilize and store temporarily at 4 °C (*see* **Note 6**). For long-term storage, aliquot and store at −20 °C for up to 6 months.

5. Gelatin coating solution 0.1 % (wt/vol): add 0.5 g of tissue culture-grade gelatin into 100 mL of PBS. Autoclave and store at 4 °C.

3   Methods (Fig. 1)

Maintain cell culture at 37 °C in a humidified incubator containing 5 % CO_2 in air. Perform all procedures under a sterilized laminar flow hood (Biological Safety Cabinet Class II). Warm up all mediums to 37 °C in a water bath. Change medium daily on mouse PSC cultures and every 2 days during the PSC differentiation stages.

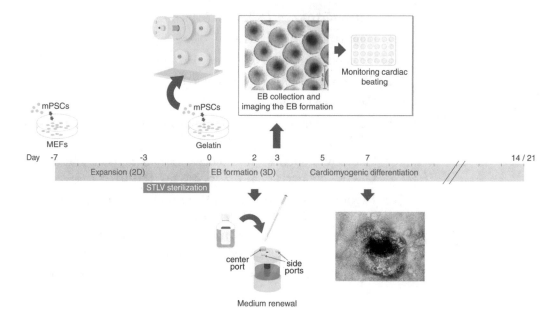

Fig. 1 Schematic illustration of the scalable cardiomyogenic differentiation from PSCs-derived EBs in a STLV bioreactor. PSCs are expanded as monolayers (2D) in 100-mm petri dishes until you obtain $\geq 3.3 \times 10^7$ cells/ one vessels. Then, PSCs are passaged and seeded in STLV vessel for generating EB formation. Medium renewal is performed on day 2 and EBs are collected on day 3 for imaging the EB formation and monitoring cardiac beating by plated EBs into 24-well plate. For cardiomyogenic differentiation, beating areas are first observed 4 days after plating and continuously monitored the beating activities on a daily basis for a period of 14–21 days. Scale bar: 500 μm

3.1 Mouse PSC Lines Preparation (Day-7–Day 0)

1. Culture mouse PSCs on feeder cells for at least two passages after thawing in complete PSC medium (LIF 1000 U/mL).

2. Culture mouse PSCs for at least one passage in feeder-free conditions on gelatin-coated 100 mm plates in complete PSC medium (LIF 2000 U/mL) (*see* **Note 7**).

3.2 STLV Bioreactor System Preparation and Sterilization (Day-3)

1. Unscrew and remove the central screws of the STLV by using an Allen wrench.

2. Separate front plate and back plate of STLV. Take of a center port cap and two side port caps.

3. Place all pieces of STLV in plastic or steel bucket for soaking with a labware detergent.

4. Scour plastic parts (both plates of vessel, a center port cap, and two side port caps) with a sponge to remove any remainders.

5. Clean silicone rubber oxygenator transfer membrane which is located along the center axis of the vessel with the tip of your finger using laboratory gloves.

6. Rinse all pieces of STLV thoroughly under continuous flow of distilled water for 20 min and soak them in distilled water

overnight. Then, remove them from distilled water and allow them to dry in laminar flow.

7. Assemble all individual parts of the STLV except a center port cap.

8. With a 25 mL pipette, fill the 110 mL STLV with 70 % ethanol through a center port to soak for 24 h.

9. Open a center port cap and two side port caps of STLV to pour 70 % ethanol.

10. Remove both a center port cap and two side port caps. Then autoclave them separately from the vessel. Cover all port caps with aluminum foil before autoclaving.

11. Loosen the central screws of the STLV (1 turn) by using an Allen wrench.

12. Cover the STLV with aluminum foil and autoclave at 121 °C for 20 min.

13. Remove the STLV and all port caps from autoclave and allow them cool down at room temperature. Keep STLV in a sterile location before use.

14. The rotator base should be cleaned with an alcohol pad or embedded cloth at least 3 times before use. When not in use, rotator should be kept in a sterile place (*see* **Note 8**).

3.3 Culture Dish and Plate Preparation

1. A gelatin-coated 24-well plate and 100-mm dish; add 0.5 and 2 mL of 0.1 % gelatin to a 24-well plate and 100-mm dish, respectively.

2. Rotate plate to evenly distribute the gelatin and cover the entire surface.

3. Place gelatin-coated plate under the laminar flow hood for 30 min.

4. Aspirate off the remaining gelatin solution.

5. Add specific medium such as the incomplete PSC medium or complete PSC medium or cardiac differentiation medium.

3.4 EB Formation in the STLV Bioreactor (Day 0)

3.4.1 Passaging Mouse PSC Lines

1. Aspirate the culture medium and rinse cells once with 5 mL of PBS 1×.

2. Aspirate PBS and add 3 mL of 0.05 % trypsin/EDTA into the plate and place them in the humidified incubator approximately 3–5 min until cell detachment occurs.

3. Gently pipette media solution with cells to further dissociate the cells mechanically.

4. Place the dissociated cells and solution into a 15 mL tube containing 3 mL of incomplete PSC medium. Spin down the cells at 1000 rpm and 4 °C for 5 min by centrifugation.

5. Discard the supernatant and resuspend cells in 10 mL of incomplete PSC medium.

6. Calculate viable cells/mL by trypan blue staining using a hemocytometer.

7. Dilute the cells using incomplete PSC medium so that the final concentration is 3×10^5 cells/mL or 3.3×10^7 cells/110 mL.

3.4.2 Seeding mPSC Lines in STLV Bioreactor

1. Unwrap STLV and gently tighten central screws in the laminar flow.

2. Fill the 90 mL STLV with incomplete PSC medium through a center port using a 25 mL serological pipette.

3. Seed the resuspended cells with incomplete PSC medium through a center port using a 10 mL serological pipette and close all port caps. Wait 5 min to allow cells to settle.

4. Open the two side port caps and screw on pre-sterilized one-way stopcocks to the two side ports.

5. Clean the side ports with an alcohol pad. Fill a 5 mL sterile syringe with incomplete PSC medium and attach this syringe to a side port using a one-way stopcock. Attach an empty 5 mL syringe to another side port using a one-way stopcock.

6. Gently and slowly move the vessel to expel air bubbles from the ports. Gently suction air bubbles by pulling the plunger of the empty syringe and gently replace air bubbles by pressing the plunger of the syringe-containing medium (*see* **Note 9**).

7. After removing air bubbles, close all stopcocks and discard all syringes. Clean the side ports with an alcohol pad and cover the side port caps.

8. Wipe any spilled media immediately from the STLV vessel with an alcohol pad to avoid contamination.

9. Connect the STLV vessel to the rotator base and incubate them inside the 5 % CO_2 incubator.

10. Connect the rotator base to the power supply box, which can be brought out of the incubator to connect to an electrical power outlet (*see* **Note 10**).

11. Set the STLV rotation speed at a speed of 10 rpm and rotate for a minimum of 24 h to monitor for potential leaks (*see* **Note 11**).

3.5 Medium Renewal (Day 2)

1. Turn off bioreactor power and remove the STLV from the rotation base and take it to a laminar flow hood.

2. Let the cells or EBs settle at the bottom for 3 min and open a side port cap and gently aspirate 50 % of medium thought this port (*see* **Note 12**).

3. Fill the 50 mL volume of incomplete PSC medium into STLV through a side port using a 25 mL serological pipette. Wait for 3 min for allowing cells to settle and follow refill medium as described above Section 3.4.2 (**steps 5–12**).

3.6 Cardiomyogenic Differentiation (Day 3–21)

1. Turn off the bioreactor power switch and remove the STLV from the rotation base and take it to a laminar flow hood.

2. Let the cells or EBs settle at the bottom for 3 min and open a side port cap and aspirate 75 % of medium thought this port.

3. For imaging the EB formation stage: gently rotate the STLV and continue to transfer the EBs to a 60-mm dish using a 1000 μL sterile pipette tip (cut ~2 cm from the extremity). Acquire the EBs images by using an Olympus IX71 inverted microscope with an attached Olympus DP 70 digital camera (*see* **Note 13**).

4. For monitoring cardiac beating: gently ·rotate the STLV and continue transfer the EBs to a 150-mm bacteriological petri dish by using a 25 mL pipette. Carefully observe and select homogenous EBs under stereomicroscope. Place individual EBs into a gelatin-coated 24-well plate containing 500 μL of cardiac differentiation medium. Continuously observe and record the beating activities on a daily basis under an inverted phase-contrast microscope for a period of 14–21 days (*see* **Note 14**). Change medium every 2nd day.

4 Notes

1. The methodology described above was optimized using an HM1 ESC line (deficient in hypoxanthine phosphoribosyl transferase ESC line [9, 10]), A4-iPSC line and B5-iPSC line (mouse Sleeping Beauty transposon-generated iPSC lines [13]). When another PSC line is utilized, the efficiency of cardiac differentiation may be affected and might need further optimization.

2. The same lot of FBS should be used during the complete series of experiments. Changing the FBS may alter the reproducibility of the EB formation and cardiac differentiation.

3. The STLV, RCCS-4H is composed of four culture vessels, a rotator base and a power supply box. The vessels are available in 55, 110, 250, 450, 500, and 1000 mL capacities. We usually use RCCS-4H with 110 mL of culture vessels, which is able to produce approximately 20,000 EBs per vessel [10]. The high capacity vessels require large volume of media, leading to laborious and expensive process. We suggest the use of RCCS-4H with 55 mL or 110 mL of culture vessels.

4. A minimum 500 mL volume of incomplete PSC medium should be prepared for the entire duration of one replication (one 110 mL-culture vessel). To prepare complete PSC medium just add LIF to incomplete PSC medium (basal medium). To prepare cardiac differentiation medium add L-ascorbic acid instead.

5. Culture mouse PSCs in feeder-free conditions. The MEFs feeder cells can be substituted by the addition of LIF supplements in the culture medium (since LIF is secreted by MEFs).

6. L-ascorbic acid should be added in cardiac differentiation medium at the working concentration before use because it may lose efficacy.

7. The mouse PSCs should be expanded until you obtain $\geq 3.3 \times 10^7$ cells/one for 110 mL-culture vessels.

8. Do not autoclave the rotator base. Also, the rotator base should not be placed inside an incubator while not in use since it will rust the rotor/propeller.

9. When replacing air bubbles with medium, do not forcefully pull back the plunger of the empty syringe and push the other syringe-containing medium. Otherwise, this process can damage the oxygenation membrane, leading to poor oxygen diffusion.

10. When the STLV power box is placed inside an incubator it can get damaged. It should be conveniently placed outside, on the top or on the side of the external wall of the incubator.

11. The rotational movements should not be interrupted except when changing medium or collecting EBs; otherwise, it will cause agglomeration of EBs.

12. While changing medium, it is easy to lose floating EBs; thus, do not shake the STLV vessel during that procedure.

13. To make a better image of EBs, move a 60-mm dish anticlockwise until all EBs are located in center. Next, move EBs to get them closer to each other by using a tuberculin syringe.

14. After plating individual EBs into gelatin-coated 24-well plates, EBs should not be disturbed for 48 h (day 4–5 of culture) to allow EBs to attach the plate. Cardiomyogenic differentiation of ESCs and iPSCs should be monitored closely by assessing the beating activity for 14 and 21 days, respectively. The cardiac differentiation efficiency from mouse iPSCs is low and delayed compared with mouse ESCs. This is supported by data showing the percentage of beating EBs and the expression of cardiac-specific messenger RNA (mRNA) for troponin [14].

Acknowledgements

The research that supports these methodologies was funded by grants from the EU FP7 ("PartnErS" PIAP-GA-2008-218205; "AniStem," PIAP-GA-2011286264; "EpiHealth," HEALTH-2012-F2-278418; "EpiHealthNet," PITN-GA-2012-317146, "STEMMAD," PIAPP-GA-2012-324451) Research Center of Excellence 9878/2015/FEKUT project, the Mahidol University, the Thailand Research Fund (TRF), the Office of Higher Education Commission, Thailand (OHEC), and the Mahidol University (MRG 5680108).

References

1. Chong JJ, Yang X, Don CW, Minami E, Liu YW, Weyers JJ, Mahoney WM, Van Biber B, Cook SM, Palpant NJ, Gantz JA, Fugate JA, Muskheli V, Gough GM, Vogel KW, Astley CA, Hotchkiss CE, Baldessari A, Pabon L, Reinecke H, Gill EA, Nelson V, Kiem HP, Laflamme MA, Murry CE (2014) Human embryonic-stem-cell-derived cardiomyocytes regenerate non-human primate hearts. Nature 510 (7504):273–277. doi:10.1038/nature13233

2. Masumoto H, Ikuno T, Takeda M, Fukushima H, Marui A, Katayama S, Shimizu T, Ikeda T, Okano T, Sakata R, Yamashita JK (2014) Human iPS cell-engineered cardiac tissue sheets with cardiomyocytes and vascular cells for cardiac regeneration. Sci Rep 4:6716. doi:10.1038/srep06716

3. Kempf H, Kropp C, Olmer R, Martin U, Zweigerdt R (2015) Cardiac differentiation of human pluripotent stem cells in scalable suspension culture. Nat Protoc 10(9):1345–1361. doi:10.1038/nprot.2015.089

4. Höpfl G, Gassmann M, Desbaillets I (2004) Differentiating embryonic stem cells into embryoid bodies. Methods Mol Biol 254:79–98. doi:10.1385/1-59259-741-6:079

5. Chen M, Lin YQ, Xie SL, Wu HF, Wang JF (2011) Enrichment of cardiac differentiation of mouse embryonic stem cells by optimizing the hanging drop method. Biotechnol Lett 33 (4):853–858. doi:10.1007/s10529-010-0494-3

6. Rungarunlert S, Techakumphu M, Pirity MK, Dinnyes A (2009) Embryoid body formation from embryonic and induced pluripotent stem cells: Benefits of bioreactors. World J Stem Cells 1(1):11–21. doi:10.4252/wjsc.v1.i1.11

7. Barzegari A, Saei AA (2012) An update to space biomedical research: tissue engineering in microgravity bioreactors. Bioimpacts 2 (1):23–32. doi:10.5681/bi.2012.003

8. Pettinato G, Wen X, Zhang N (2015) Engineering strategies for the formation of embryoid bodies from human pluripotent stem cells. Stem Cells Dev 24(14):1595–1609. doi:10.1089/scd.2014.0427

9. Rungarunlert S, Klincumhom N, Bock I, Nemes C, Techakumphu M, Pirity MK, Dinnyes A (2011) Enhanced cardiac differentiation of mouse embryonic stem cells by use of the slow-turning, lateral vessel (STLV) bioreactor. Biotechnol Lett 33(8):1565–1573. doi:10.1007/s10529-011-0614-8

10. Rungarunlert S, Klincumhom N, Tharasanit T, Techakumphu M, Pirity MK, Dinnyes A (2013) Slow turning lateral vessel bioreactor improves embryoid body formation and cardiogenic differentiation of mouse embryonic stem cells. Cell Reprogram 15(5):443–458. doi:10.1089/cell.2012.0082

11. Lü S, Liu S, He W, Duan C, Li Y, Liu Z, Zhang Y, Hao T, Wang Y, Li D, Wang C, Gao S (2008) Bioreactor cultivation enhances NTEB formation and differentiation of NTES cells into cardiomyocytes. Cloning Stem Cells 10 (3):363–370. doi:10.1089/clo.2007.0093

12. Yirme G, Amit M, Laevsky I, Osenberg S, Itskovitz-Eldor J (2008) Establishing a dynamic process for the formation, propagation, and differentiation of human embryoid bodies. Stem Cells Dev 17(6):1227–1241. doi:10.1089/scd.2007.0272

13. Muenthaisong S, Ujhelly O, Polgar Z, Varga E, Ivics Z, Pirity MK, Dinnyes A (2012) Generation of mouse induced pluripotent stem cells

from different genetic backgrounds using Sleeping beauty transposon mediated gene transfer. Exp Cell Res 318(19):2482–2489. doi:10.1016/j.yexcr.2012.07.014

14. Mauritz C, Schwanke K, Reppel M, Neef S, Katsirntaki K, Maier LS, Nguemo F, Menke S, Haustein M, Hescheler J, Hasenfuss G, Martin U (2008) Generation of functional murine cardiac myocytes from induced pluripotent stem cells. Circulation 118(5):507–517. doi:10. 1161/CIRCULATIONAHA.108.778795

Methods in Molecular Biology (2016) 1502: 181–194
DOI 10.1007/7651_2015_317
© Springer Science+Business Media New York 2015
Published online: 12 February 2016

Whole-Heart Construct Cultivation Under 3D Mechanical Stimulation of the Left Ventricle

Jörn Hülsmann, Hug Aubin, Alexander Wehrmann, Alexander Jenke, Artur Lichtenberg, and Payam Akhyari

Abstract

Today the concept of Whole-Heart Tissue Engineering represents one of the most promising approaches to the challenge of synthesizing functional myocardial tissue. At the current state of scientific and technological knowledge it is a principal task to transfer findings of several existing and widely investigated models to the process of whole-organ tissue engineering. Hereby, we present the first bioreactor system that allows the integrated 3D biomechanical stimulation of a whole-heart construct while allowing for simultaneous controlled perfusion of the coronary system.

Keywords: Whole-organ tissue engineering, Whole heart bioreactor, Biomechanical stimulation, 3D stretching, Myocardial tissue engineering

1 Introduction

In the concept of *Whole-Organ Engineering* entire donor organs are decellularized and used as a three-dimensional scaffold with an intact macro- and micro-architecture as well as a vascular system. As this is a complex and rather holistic approach, a composed adjustment of controlling chemical and biological reaction-kinetic parameters represents a keystone, concerning mainly mass transport, cell growth, and stimulation of tissue maturation. Thereby the main functions of a whole-organ bioreactor are:

1. *Pressure controlled perfusion of the organ's vascular system*
   (this enables the transport of metabolites into and out of the depth of the organ where they can reach stroma and parenchyma by diffusion).
2. *Processing of cell culture medium*
   (in terms of concentrations of various metabolites and other ingredients, pH, and temperature).
3. *Integration of stimulation approaches for guiding tissue maturation*
   (biochemical and biophysical stimulation).

4. *Online detection for feedback control*
(monitoring of critical process-parameters, e.g., the biophysical
stimulation intensity).

When the complex nature of myocardial tissue is broken down
to measurable biophysical functions, the three-dimensional electro-
dynamical conductivity is one of the main remaining principles
besides mechanical forces. Respectively, the natural function of
myocardium strongly relies on the unique cellular spatial orientation and functional integrity throughout the myocardial tissue.
This is strongly supported by findings from previous studies involving in vitro stretching of myocardial constructs [1–4]. The herein
described bioreactor system could transfer the concept of
biomechanical stimulation from artificially dimensioned constructs
to the natural proportions of a whole-heart scaffold with its aniso-
tropic three-dimensional ECM and the superior spatial design of
the left ventricle. In previous studies we have shown that although
stimulated whole-heart constructs developed less measurable cellular viability in contrast to non-stimulated constructs, they revealed a
clearly increased cellular orientation and arrangement [5].

As an alternative to biomechanical stimulation, myocardial
tissue can also be stimulated by exposure to an electrophysiological
environment through the application of electric tension or direct
current injection [6, 7]. However, only direct current injection has
been previously investigated in a setting involving whole-heart
constructs [8]. For artificial tissue constructs of less complexity
the combination of mechanical and electrophysiological stimulation has already revealed an additional benefit [9]. Hence, it is a
logical prospective for the future to transfer this concept of dual or
multimodal stimulation to the whole-heart bioreactor as well.
Congruously, this points clearly to another demand of bioreactors
in Tissue Engineering in general—the open modularity in an
integrated control system, which allows the system to integrate
further devices and to advance with ongoing scientific progress.
Therefore, for the presented bioreactor system we have applied
LabVIEW® as complete process control system (PCS), providing
fully integrated control of performed cultivations [5].

The main challenge for a whole-heart bioreactor is to unify and
transfer the findings of fundamental myocardial tissue engineering
in an as much holistic and nature like approach as possible. At the
current state of science and technology this means rather to screen
for an optimal stimulation and cultivation procedure in a system
with a suitable exemplary cell type. Therefore, cells should be able
to respond to applied stimuli (e.g., micro-architecture and mechanical stretching) and have the capability to mature to cardiomyocytes. Hence, neonatal cardiomyocytes represent a widely used cell
type in myocardial tissue engineering applications as they can be
easily isolated in large numbers and because they are predestinated

for maturation to adult cardiomyocytes [9]. However, this maturation process may be disturbed in inadequate settings, as their cultivation in monolayers typically results in dedifferentiation which is visible morphologically as well as at a functional and molecular level [10]. In contrast, terminal differentiation as well as organotypic assembly and maturation of neonatal cardiomyocytes is successfully induced within 2 weeks using 3D-engineered heart tissue (EHT) constructs subjected to phasic mechanical stretch [11]. Thus, prospective repopulations and cultivations with clinically relevant cell types should be based on fundamental knowledge concerning optimized bioprocessing strategies for whole-heart constructs. However, the demands for an optimal cell type with clinical relevance are still challenging up to today [12]. Although some progress has been achieved and recent work has shown significant formation of myocardial functionality for human iPSC-derived cardiac myocytes in human whole-heart constructs [13], the optimal cell types for clinically relevant whole-heart tissue engineering are still under debate and need further evaluation in the future.

2 Materials

2.1 Recellularization of the Coronary Vascular System

2.1.1 Recellularization

1. Sterile cloth.
2. Sterile gloves.
3. Six-well.
4. 50 ml falcon.
5. 5 ml syringe.
6. 2 ml syringe.
7. Needle (for 5 ml syringe).
8. Closing stopper.
9. Sterile tweezers.

2.1.2 Solutions

1. Phosphate buffered Saline.
2. Penicillin–streptomycin. Storage at $-20\,^{\circ}$ C.
3. Dulbecco's Modified Eagle Medium (DMEM) with 10 % fetal calf serum.
4. Endothelial Cell Growth Medium (ECGM).

2.1.3 Cell Population

1. Endothelial cells (HUVECs = Human Umbilical Vein Endothelial Cells) (2.5 Mio cells in total, in 5 ml cell suspension).

2.2 Recellularization of the Stroma

2.2.1 Recellularization

1. Sterile cloth.
2. Sterile gloves.
3. Six-well.
4. 50 ml falcon.

5. 1 ml syringe.

6. 2 ml syringe.

7. 5 ml syringe.

8. 24 gauge needle (for 1 ml syringe).

9. Closing stopper.

10. Sterile tweezers.

2.2.2 Solutions

1. Phosphate buffered Saline.

2. Penicillin–streptomycin. Storage at −20 ° C.

3. Dulbecco's Modified Eagle Medium (DMEM) with 10 % fetal calf serum.

2.2.3 Cell Population

1. Neonatal cardiomyocytes (5 Mio cells in total, in 1 ml of cell suspension) from neonatal rats.

2.3 Bioreactor Hardware

2.3.1 Processing Flask (as Depicted in Fig. 1)

1. Modified Schott Duran square cut flask, 250 ml (2 added tube clips).

2. Screw Cap GL 45, 3 Port GL 14.

3. GL 14 inset 1.3 mm for PTFE tubing (stimulation System).

4. GL 14 inset 6 mm for silicone tubing (construct Perfusion).

5. 0.2 μm membrane filter cap for pressure compensation.

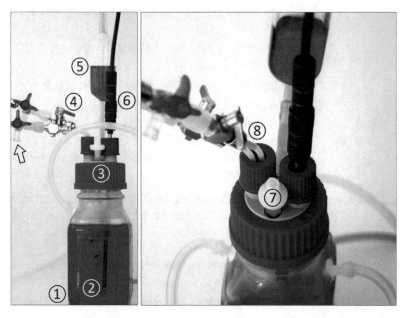

Fig. 1 Processing flask. In detail: (*1*) 250 ml Schott Duran flask; (*2*) planar glass pane inlet for optical access; (*3*) Screw cap GL 45; (*4*) Stopcocks for stimulation pressure pipe; (*5*) bubble trap for blister-free perfusion; (*6*) PT 100 temperature sensor; (*7*) 0.2 μm membrane filter for pressure compensation; (*8*) PTFE pressure tubing inlet

6. Glass bubble trap—bulgy glass pipe.

7. MU-Pt-100-U010 temperature sensor (Suran Industrieelektronik, Horb, Germany).

2.3.2 Tubing System

1. Silicone Tube 3×1.6 mm.

2. Luer-Lock connector system male/female for 3.2–4.2 mm.

2.3.3 Stimulation System

1. Tecan XLP 6000 syringe pump (Tecan, Männedorf, Switzerland).

2. Sera RS 204.1-1.2e diaphragm pump (Seybert und Rahier, Immenhausen, Germany).

3. Custom made connector from pump thread to 1.3 mm PTFE tubing.

4. JUMO dTRANS p30 pressure transducer (Jumo, Fulda, Germany).

5. PTFE Tubing 1.5×3.2 mm.

6. PTFE Multifit T-connectors (Reichelt Chemietechnik, Heidelberg, Germany).

7. PTFE hollow screws for 3.2 mm bore (Reichelt Chemietechnik, Heidelberg, Germany).

8. Metal Luer-Lok stopcock connector, autoclavable (Neolab, Heidelberg, Germany).

9. Metal tubeclip with Luer-Lok cone (Neolab, Heidelberg, Germany).

10. Left ventricle pressure balloons (Radnoti LCC, California, USA), sizes 4 and 5.

11. Sealing compound (Dow Corning, Wiesbaden, Germany).

12. Reservoir : Sterile D-PBS 0.5 l in original manufacturer GL 45 flask (Hoffmann-La Roche AG, Basel, Switzerland) connected via GL 45 screw cap—3 Port GL 14 and GL 14 inset 1.3 mm for PTFE tubing. Additional 0.2 μm membrane filter cap for pressure compensation.

The assembled stimulation system in use with a mounted whole heart, particularly the sealed end of the pressure pipe with the ligated left ventricle balloon, is illustrated in Fig. 2.

2.3.4 Perfusion System

1. Medex™ logical pressure transducer (Smiths Medical, St. Paul, USA).

2. Medex™ pressure cartridge (Smiths Medical, St. Paul, USA).

3. Kinesis Reglo ICC peristaltic pump (Kinesis GmbH, Langenfeld, Germany).

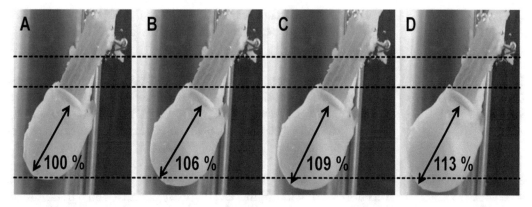

Fig. 2 Whole ventricle stretching. The figure displays a WHS mounted to the combined perfusion and stimulation system at varying stretching conditions in the sequence from A to D. The *upper dashed line* marks a prominent fix area at the sealed end of the pressure pipe. The *middle dashed line* marks the upper end of the ventricle-balloon as another fix point. The *lower dashed line* crosses the tip of the WHS's apex at the unstrained condition at A. The *double arrows* mark the longitudinal length of the left ventricle wall from base to apex at a planar vision. In the discrete sequence from A to D a volume-stroke is depicted that results in a direct longitudinal stretching of 13 % under negligence of the curvature

2.3.5 Process Control/ Field Control Level	Analogue/digital converter: Advantech ADAM 4017 analogue input module/ADAM 4024 analogue output module/ADAM 4520 serial converter/Advantech USB 4716 data acquisition module (Advantech, New Taipei City, Taiwan).
2.3.6 Process Control System	LabVIEW® Developer Suite including the control design and simulation option (National Instruments, Austin, USA).
2.4 Cellular Viability Assay	WST-1 proliferation reagent® (Roche Diagnostics GmbH, Mannheim, Germany, REF: 05015944001).
2.5 Staining of Nucleic Acids and Cytoskeleton	1. Acridine Orange Staining Solution (AO) (Molecular Probes, cat.nr. A1301). 2. Phalloidin/Rhodamine conjugate (Invitrogen, Carlsbad, California, USA, cat. no. A12379).

3 Methods

3.1 Acellular Whole-Heart Scaffold

For the production of acellular whole-heart scaffolds (WHS) rodent hearts from adult male Wistar rats (400–500 g) were decellularized via automated software-controlled coronary perfusion with standard decellularization detergents preserving the basic anatomy, vascular network, and critical ECM characteristics of the native heart, as previously published [14]. Automation of the decellularization process allows standardization of produced scaffolds. Published protocols can be adapted to human-sized donor hearts. Acellular heart scaffolds are stored in PBS supplemented with 100 IU/ml penicillin–streptomycin at 4 °C until further processing.

3.2 Repopulation of the Whole-Heart Scaffold

WHS are repopulated by a combined method of reseeding both the vascular system and the parenchyma. Thereby the scaffolds turn to whole-heart constructs (WHC). The WHC are allowed to solidify cellular attachment overnight prior to mounting into the bioreactor for cultivation.

3.2.1 Recellularization of the Coronary Vascular System

The intact coronary vascular system of the acellular WHS can be recellularized with high selectivity by manual retrograde aortic perfusion with human umbilical vein endothelial cells (HUVEC). Two and a half million cells are needed for complete re-endothelialization of the coronary vascular system of a rodent WHS. Following protocol can be adopted to other endothelial cell types and donor scaffolds, cell number for complete re-endothelialization may vary. Concomitantly please consider notes 1–3.

1. The following steps should be performed under a laminar flow bench under sterile conditions. PBS and cell culture medium should be pre-warmed to 37 °C.

2. Place the WHS in a six-well with DMEM.

3. Pre-perfuse the WHS with 5 ml DMEM from a 5 ml syringe.

4. Prepare a 5 ml cell suspension of 5×10^5 HUVECs per ml in ECGM.

5. Perfuse the vascular system with 5 ml of the cell suspension.

6. Place the re-endothelialized WHC into a 50 ml falcon with 15 ml of ECGM. Leave open the cover of the falcon slightly to allow circulation of oxygen. Store the falcon in an incubator at 37 °C and 5 % CO_2.

7. Wait for 4 h to allow cell attachment of the endothelial cells to the vascular system, then cautiously perfuse the coronary system with 5 ml of fresh ECGM medium to wash out non-attached and nonviable cells. Change the medium in the falcon tube as well.

8. Incubate the WHC in a 6-well while preparing parenchymal reseeding according to Section 3.2.2.

3.2.2 Recellularization of the Stroma

The stroma or interstitial or parenchymal area of the left ventricular free wall of the acellular whole-heart scaffold can be recellularized through manual interstitial injection with rodent neonatal cardiomyocytes. Five million neonatal cardiomyocytes are isolated and prepared following standard protocols. Following protocol can be adopted to other cell types and donor scaffolds, cell number for

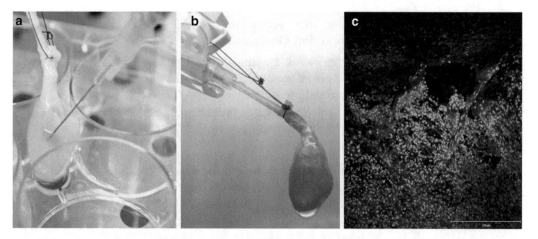

Fig. 3 Seeding and analytical evaluation of WHC. (**a**) WHS are seeded by injection of cell suspension into the left ventricle wall. The contrast of the red colored medium to the pale WHS is clearly recognizable at the injection site. (**b**) A repopulated WHC after perfusion and incubation of WST-1 cellular viability reagent. The colorimetric conversion of the reagent visualizes areas of high cellular viability inside of the WHC. (**c**) Confocal microscopy of a repopulated and cultivated WHC after AO staining. This insight reveals a widespread 3D parenchymal repopulation. Furthermore, based on the obtained graphical impression the overall cellular orientation is adjusted to the micro-architecture of the ECM

standardized parenchymal recellularization may vary. An illustration of the recellularization technique by injection is shown in Fig. 3a. Concomitantly please consider notes 1, 3, and 4.

1. Following steps should be performed under a laminar flow bench under sterile conditions. PBS and cell culture medium should be pre-warmed to 37 °C.

2. Place the WHC in a six-well with PBS.

3. Pre-perfuse the WHC with 5 ml PBS from a 5 ml syringe to completely wash-out cell culture medium, in case that the vascular system has been recellularized beforehand.

4. Manually perfuse the WHC with PBS from a 5 ml syringe to build up pressure and balloon the heart.

5. Place the ballooned WHC into an empty six-well, so that the free anterior wall of the left ventricle faces upside.

6. Prepare a 1 ml cell suspension of 5×10^6 neonatal cardiomyocytes in DMEM in a 1 ml syringe equipped with a 24 gauge needle.

7. Start the interstitial recellularization by slowly injecting 0.1 ml of the cell suspension into the left ventricular interstitium. Take special caution to not perforate the ventricular wall.

8. This step is being repeated 10× until a total volume of 1 ml is injected.

9. Perfuse the recellularized WHS with 5 ml DMEM to re-condition the stroma.

10. Fill the cannula with 2 ml of DMEM from a 2 ml syringe to avoid air trapped in the cannula. Put a closing stopper on the cannula of the heart.

11. Place the recellularized WHS into a 50 ml falcon with 15 ml of DMEM. Leave open the cover of the falcon slightly, to allow circulation of oxygen.

12. Wait for 4 h to allow cellular attachment of the cardiomyocytes to the interstitium, then cautiously perfuse the coronary system with 5 ml of fresh DMEM (ECGM in case of prior vascular recellularization) to wash out non-attached and nonviable cells. Change the medium in the falcon tube as well.

13. Incubate the WHC in a 6-well overnight prior to mounting into the bioreactor. If cellular viability should be determined prior to bioreactor cultivation please refer to Section 3.5.

3.3 Preparation of the Hardware System

The assembly of the tubing system and the processing flask together with the stretching system has to be done in the following order prior to autoclaving the whole system in an autoclave-bag.

1. Prepare a short piece of PTFE Tube (about 20 cm length) with a rasped round notch of about 0.5 mm depth and 1 mm width at 3–5 mm to the end. For sealing this end has to be overdrawn thinly with silicone and let be hardened overnight. Connect the other end to a metal tube clip with Luer-Lock cone and a metal Luer-Lock Stopcock connector to enable filling and closing.

2. Prefill the sealed PTFE Tube and an appropriate latex balloon (size 0.06–0.2 ml for rat hearts) with PBS. Then slip the balloon over the sealed end and ligate it tightly into the notch. Close the stopcock before autoclave.

3. Put together the silicone tubing and close the bottle top of the processing vessel. Cover the connection sites of the pt-100 with latex, e.g., by a cut finger tip of a lab-glove before autoclave.

4. The complete perfusion system (processing vessel + tubing) system should be autoclaved in an autoclave bag and put into a sterile bench for prefilling.

5. Prefill the perfusion system with warm cell culture medium. The attachment tubes for the perfusion-pressure transducer should be connected and prefilled sterilely with medium while prefilling the tubing system.

6. Perfuse the seeded WHC with medium and insert the balloon into the left ventricle. Then mount the Luer-Lock end of the cannula to the perfusion connection. For sterile handling use autoclaved tweezers.

7. Mount the appropriate tubing sections into the perfusion pumps and attach the attachment tubes to the perfusion-pressure transducer.

8. Start the perfusion manually with constant flow at low perfusion pressure.

9. Prefill the PTFE pressure pipes of the stimulation system by flushing them via the TECAN XLP 6000, thereby remove all air. Simultaneously start the SERA pump at low stroke and frequency to vent into the tubing system.

10. After removing all air, attach the stimulation system to the ballooned pipe and the pressure transducer. Then vent the system again to the side of the pressure transducer by flushing the pipe system.

3.4 Setting the Control System

Both perfusion pressure and the amplitude of the intrinsic pressure of the stimulation system have to be controlled over the period of cultivation according the desired experimental design. The concept of control is adapted to the characteristics of the perfusion loop by a classic PI algorithm (proportional integral) and for the stimulation system by a modified 2-point control with a converging function. To modulate each cultivation to the individual WHC a perfusion step function has to be performed at the start to estimate the individual regulator parameters. For the stimulation system only the pressure-to-volume ratio has to be set according to the desired stretching intensity corresponding to the pressure to volume relation. The modulation runs automatically by the converging modus. Concomitantly please consider notes 5 and 6.

3.4.1 Setting of Stimulation Control

1. Charge the balloon discrete with small volume increments and record the resulting intrinsic static pressure.

2. Observe the resulting longitudinal stretching graphically, e.g., via digital photography. Measure or calculate the expansion. An example for a graphical evaluation is given in Fig. 2.

3. Determine the differential quotient of the pressure-to-volume relation at a range accordant to the aspired longitudinal distension of the WHC.

4. Set the determined differential quotient as parameter to the 2-point control interface.

5. Use the converging function appropriate to the intensity and frequency of stretching and the aspired duration of continuous stimulation.

6. Prefill the balloon to the aspired volume using the XLP 6000 control interface.

7. Set the aspired stimulation frequency to the stimulation interface.

8. Set the aspired stroke intensity manually to the control knob at the front side of the SERA Pump.

9. Activate the stimulation system.

10. Wait at least for five strokes and activate stimulation control.

3.4.2 Setting of the Perfusion Control

1. Adjust the revolution speed and the occlusivity so that a perfusion pressure of about 150 mmHg can be effectuated by a revolution speed of about 20–50 rpm. Then set the pump to 0 rpm, execute the step function, and record the step response. Determine the delay (T_u) and the equalization time (T_m) according to the inflectional tangent principle. The stationary gain (K_s) equates to set rpm at the step function. To calculate the controller gain (K_c) and integration time (T_i) for aperiodic standard course use following formula:

$$K_c = \frac{0.35}{K_s} \cdot \frac{T_m}{T_u}$$

$$T_i = 1.2 \cdot T_m$$

2. Set the parameters to the software and start the feedback control.

3.5 Estimation of Cellular Viability

The development of the WHC's cellular viability can be investigated by the application of the non-destructive and repetitive WST-1 proliferation reagent®. We adapted the standard protocol as supplied by the manufacturer to the whole-organ concept by standardizing a perfusion-saturation setting. Thereby, the working concentration of the WST-1 reagent should be adapted to the used cell type beforehand. For instance constructs could be subjected to this viability assay. A WHC that has been perfused and incubated with WST-1 reagent is depicted in Fig. 3b.

1. Prepare working solution of WST-1 reagent in the particular cell culture medium as used for the cultivation.

2. Perfuse 5 ml pre-warmed PBS to the WHC. Use low perfusion force maintaining evenly dripping of the construct, avoiding continuous flow.

3. Perfuse 5 ml pre-warmed WST-1 working solution to the WHC. Use low perfusion force maintaining evenly dripping of the construct, avoiding continuous flow.

4. Incubate the WHC in an empty 6-well. Place the WHC in one of the middle wells and the cannula site into an outer well. The connecting cannula with the ligated aorta then bends over the bordering walls of the wells. Incubate up to 4 h in a CO_2 Incubator at appropriate conditions for the cells and the used medium.

5. After incubation elute the reagent by perfusing the WHC with 1 ml of fresh medium and collect the perfusate.

6. Measure the absorbance of the formazan product at 440 nm against 600 nm as reference wavelength. Use pure working solution as blank control.

3.6 Determination of Repopulation and Cellular Constitution in the ECM

The spatial repopulation of the ECM and the cellular spreading in the WHC can be investigated simply by confocal microscopy. Staining nucleic acids by Acridine Orange (AO) enables the visualization of nuclei via its DNA as well as the cellular body via its RNA. Additional phalloidin staining of the WHC provides the visualization of thy cytoskeleton, what serves to determine the cells spatial spreading.

A confocal microscopy of a widespread united cell structure inside of a WHC is shown in Fig. 3c.

1. Prepare a working solution of 0.1 mg/ml AO in HBSS.

2. Perfuse 5 ml of warm HBSS to the cultivated WHC.

3. Perfuse 5 ml of AO working solution to the WHC.

4. Incubate the WHC in an empty 6-well. Place the WHC in one of the middle wells and the cannula site into an outer well. The connecting cannula with the ligated aorta then bends over the bordering walls of the wells. Incubate for 20 min at 37 °C—e.g., in a CO_2 Incubator.

5. Perfuse 5 ml of warm HBSS to the WHC.

6. Dissect the left ventricular wall (LV) and place it into a 6-well with 3 ml of warm HBSS.

7. Prepare a working solution of 2 U phalloidin/rhodamine conjugate (Invitrogen, Carlsbad, California, USA, cat. no. A12379) in the respective cell culture medium.

8. Place the LV into a 1.5 ml microreaction tube and add 200 μl of warm working solution.

9. Incubate the LV in the microreaction tube for 20 min at 37 °C—e.g., in a CO_2 Incubator. Preserve the preparation against exposure to light.

10. Wash the preparation for 5 min at 37 °C in warm HBSS. Preserve the preparation against exposure to light.

The double stained LV preparation then can be examined directly or subjected to fixing—e.g., in 4 % paraformaldehyde + 10 % sucrose in PBS for 30 min at 4 °C prior to examination under the confocal microscope.

3.7 Quantification of Cellular Alignment

The cellular alignment correlates to the circularity (C) and spatial orientation angle (O) of the nuclei. Referring to this, confocal microscopic pictures can be evaluated using Image J software (NIH, USA).

1. Calculate 2D distributions of O vs. C allocating the nuclei in discrete fractions for C and O—e.g., using a numerical script.

2. Identify fractions of nuclei that are clearly aligned in nearby related angle ranges at clearly straightened circularities.

3. Determine fractions that thereby remarkably differ from the bulk fractions. These can be identified as fractions of clearly oriented cells (FO).

4. Relate the cumulative count of nuclei in the FO to its corresponding angle range and thus calculate the specific orientation density (OD). The OD serves as a quality control for the level of cellular alignment in WHC.

Supposable applications of the OD would be to comparatively investigate different stretching intensities or frequencies vs. static cultivation or the capability of different cell types to align in the WHC.

4   Notes

1. While working with WHS generally leave the cannula already used for perfusion-based decellularization in the aorta to facilitate recellularization. Always keep it filled with appropriate fluid, i.e., medium or PBS and avoid the inlet of air by. Lock the cannula with a closing stopper whenever no perfusion is required.

2. While perfusing the WHS at vascular recellularization, check for ballooning of the aorta in order rule out leakage through the aortic valve and ensure sufficient coronary perfusion. Perfuse slowly and with a constant pace and pressure, so that only one drop at a time is pressed out of the apex of the heart. Pay attention to the resistance of the vascular system. Collect the spillover of the cell suspension on the bottom of the six-well to examine seeding efficiency.

3. While pre-perfusing the WHC at stromal or parenchymal recellularization perfuse the WHC with PBS until it has become translucent again, losing the coloring from the cell culture medium. Be aware that too much pressure can wash out endothelial cells in the coronary vascular system.

4. While injecting cell suspension at stroma or parenchyma reseeding an effective injection spot can be visualized by the spreading color of the culture medium around the spot. Use standardized spots for the injections, such as next to large vessel bifurcations. Collect the spillover of the cell suspension on the bottom of the six-well to examine seeding efficiency.

5. For the control circle of the stimulation pressure amplitude disturbances appear mainly by loss of fluid and gas inlet at connections of the tubing system. Regular interceptions for degassing should be planned. For cultivation with stretching frequency of more than 1 Hz and/or stroke intensities of more than 60 % and/or stretching intensities of clearly more than 10 % select a 3–5 step conversion. For other cultivations a setting of 5–10 steps for conversion is recommended.

6. Adjust the sampling rate to the stimulation frequency according to the Nyquist Theorem to avoid incorrect acquisition of the pressure amplitude. Adjust Buffering and Filter Settings to the Sampling rate of the PCS as well to avoid incorrect data acquisition.

References

1. Zimmermann W (2001) Tissue engineering of a differentiated cardiac muscle construct. Circ Res 90(2):223–230. doi:10.1161/hh0202. 103644

2. Akhyari P, Fedak PWM, Weisel RD et al (2002) Mechanical stretch regimen enhances the formation of bioengineered autologous cardiac muscle grafts. Circulation 106(12 Suppl 1): I137–I142

3. Birla R, Huang Y, Dennis R (2007) Development of a novel bioreactor for the mechanical loading of tissue-engineered heart muscle. Tissue Eng 13(9):2239–2248. doi:10.1089/ten. 2006.0359

4. Govoni M, Lotti F, Biagiotti L et al (2014) An innovative stand-alone bioreactor for the highly reproducible transfer of cyclic mechanical stretch to stem cells cultured in a 3D scaffold. J Tissue Eng Regen Med 8(10):787–793. doi:10.1002/term.1578

5. Hülsmann J, Aubin H, Kranz A et al (2013) A novel customizable modular bioreactor system for whole-heart cultivation under controlled 3D biomechanical stimulation. J Artif Organs 16(3):294–304. doi:10.1007/s10047-013-0705-5

6. Thavandiran N, Nunes SS, Xiao Y et al (2013) Topological and electrical control of cardiac differentiation and assembly. Stem Cell Res Ther 4(1):14. doi:10.1186/scrt162

7. Tandon N, Taubman A, Cimetta E et al (2013) Portable bioreactor for perfusion and electrical stimulation of engineered cardiac tissue. Conf Proc IEEE Eng Med Biol Soc 2013: 6219–6223

8. Ott HC, Matthiesen TS, Goh S et al (2008) Perfusion-decellularized matrix: using nature's platform to engineer a bioartificial heart. Nat Med 14(2):213–221. doi:10.1038/nm1684

9. Kensah G, Gruh I, Viering J et al (2011) A novel miniaturized multimodal bioreactor for continuous in situ assessment of bioartificial cardiac tissue during stimulation and maturation. Tissue Eng Part C Methods 17 (4):463–473. doi:10.1089/ten.TEC.2010. 0405

10. Tiburcy M, Didie M, Boy O et al (2011) Terminal differentiation, advanced organotypic maturation, and modeling of hypertrophic growth in engineered heart tissue. Circ Res 109(10):1105–1114. doi:10.1161/CIRCRE SAHA.111.251843

11. Tiburcy M, Meyer T, Soong PL et al (2014) Collagen-based engineered heart muscle. In: Radisic M, Black LD III (eds) Methods in molecular biology. Springer, New York, NY, pp 167–176

12. Suuronen EJ, Ruel M (eds) (2015) Biomaterials for cardiac regeneration. Springer International Publishing, Cham

13. Guyette JP, Charest J, Mills RW et al (2015) Bioengineering human Myocardium on native extracellular matrix. Circ Res: CIRCRE SAHA.115.306874. doi:10.1161/CIRCRE SAHA.115.306874

14. Aubin H, Kranz A, Hülsmann J et al (2013) Decellularized whole heart for bioartificial heart. In: Walker J (ed) Methods in molecular biology. Cellular cardiomyoplasty. Springer, Clifton, NJ

Methods in Molecular Biology (2016) 1502: 195–202
DOI 10.1007/7651_2016_332
© Springer Science+Business Media New York 2016
Published online: 10 April 2016

Tendon Differentiation on Decellularized Extracellular Matrix Under Cyclic Loading

Daniel W. Youngstrom and Jennifer G. Barrett

Abstract

Tendon bioreactors combine cells, scaffold, and mechanical stimulation to drive tissue neogenesis ex vivo. Faithful recapitulation of the native tendon microenvironment is essential for stimulating graft maturation or modeling tendon biology. As the mediator between cells and mechanical stimulation, the properties of a scaffold constitute perhaps the most essential elements in a bioreactor system. One method of achieving native scaffold properties is to process tendon allograft in a manner that removes cells without modifying structure and function: "decellularization." This chapter describes (1) production of tendon scaffolds derived from native extracellular matrix, (2) preparation of cell-laden scaffolds prior to bioreactor culture, and (3) tissue processing post-harvest for gene expression analysis. These methods may be applied for a variety of applications including graft production, cell priming prior to transplantation and basic investigations of tendon cell biology.

Keywords: Regenerative medicine, Tissue engineering, Mesenchymal stem cells, Mechanobiology, Tendon, Extracellular matrix, Decellularization, Bioreactor, Scaffold

1 Introduction

Mechanical stimulation drives tendon metabolism [1]. The elastic, collagen-rich extracellular matrix (ECM) of tendon not only transmits forces between muscle and bone [2] but also actively participates in regulating cell behavior. The combination of ECM and force contributes to tendon cell morphology [3] and differentiation state [4]. This occurs through physical coupling of cell to ECM [5], perturbation of matricellular proteins [6] and regulation of soluble growth factors [7], among other functions [8]. As a result of the reciprocal activity between ECM-synthesizing tenocytes and their products, tendons dynamically remodel in response to their specific physiological functions [9]. It is therefore no surprise that ECM is an essential component of tendon reconstruction and engineering strategies [10]. However, translating the structure and function of ECM to a tool for cell culture has remained a challenge.

Scaffold macromolecular composition [11], mechanical properties [12], baseline tension [13] and cyclic strain [14] all contribute to the mechanobiological niche of tendon cells. As a result, systems incorporating mechanical stimulation protocols [15], natural [16] or synthetic [17] extracellular matrix or combinations thereof [18] have been engineered to facilitate in vitro culture of tendon cells and their precursors. We have reviewed the progression of the tendon tissue engineering field elsewhere [19]: a collaborative and interdisciplinary process that has gradually improved the fidelity of bioreactors to mimic the structure and function of native tendon ex vivo.

Despite tremendous progress, manufacturing scaffolds that maintain not only native architecture and mechanical properties, but also proper cell-scale force transduction [20], has remained a significant challenge. To address this shortcoming, our laboratory optimized a detergent-based decellularization protocol for use with equine flexor tendons [21]. We have used decellularized tendon scaffolds (DTS) to investigate amplitude dependence of cyclic strain-induced differentiation of mesenchymal progenitor cells toward tenocytes [22], as well as to compare different cell types for potential use in cell therapy [23]. Herein we describe in detail decellularization methods, culture protocols and sample processing techniques used in our lab.

2   Materials

2.1 Tendon Decellularization

Padgett Model B Electric Dermatome: Integra LifeSciences.

Sterile surgical instruments, dermatome blades, locking clamps (stored with dermatome).

Phosphate buffered saline (PBS) ($1\times$) without calcium or magnesium: Lonza #17-516Q.

Tris base: Fisher #BP152 and hydrochloric acid: Sigma #H1758.

Sodium dodecyl sulfate (SDS): Sigma #L3771.

Multi-platform orbital shaker, placed in 4 °C refrigerator: Fisher #13-687-700.

Trypsin, 0.05 % ($1\times$) with EDTA 4Na: Gibco #25300062.

Protease inhibitor cocktail for mammalian cell and tissue extracts: Sigma #P8340.

Deoxyribonuclease (DNase) I solution (1 mg/mL): Stemcell Technologies # NC9007308.

Ethyl alcohol, 200 proof for molecular biology: Sigma #E7023.

Sterile, autoclaved, distilled water.

2.2 Bioreactor Culture

Bioreactor system: base computer, control hardware, stage, clamps, and vessels.

Low-glucose GlutaMAX DMEM with 110 µg/mL sodium pyruvate: Gibco #10567-022.

10 % MSC FBS: Gibco #12662-029.

Penicillin–streptomycin solution: Sigma #P4333.

Penicillin G sodium salt: Sigma #P3032.

L-ascorbic acid powder: Sigma #A4403.

2.3 Sample Processing

Liquid nitrogen.

RNase AWAY decontamination reagent: Molecular BioProducts #7003.

Water, sterile (for RNA work) DEPC-treated and nuclease-free: Fisher #BP561.

Guanidine thiocyanate (GITC): Fisher #BP221.

Sodium citrate, for molecular biology ≥99 %: Sigma #C8532.

Sodium hydroxide (NaOH), reagent grade ≥98 % pellets (anhydrous): Sigma #S5881

Sodium lauroyl sarcosinate (sarkosyl), for molecular biology ≥95 %: Fisher #BP234.

β-mercaptoethanol, for molecular biology 99 %: Sigma #M3148.

Sterile surgical instruments, including forceps and scoopula.

SPEX SamplePrep 6770 Freezer/Mill (cryomill).

Impactor tube, steel end caps, steel impactor set and opening tool for cryomill.

Triton X-100: Sigma #T8787.

Sorvall Legend Mach 1.6R desktop centrifuge.

Sodium acetate buffer solution, pH 5.2 ± 0.1 for molecular biology: Sigma #7899.

Saturated phenol: Ambion #AM9712.

Chloroform, contains 100–200 ppm amylenes, >99.5 %: Sigma #2432.

Sorvall RC 6 Plus superspeed centrifuge.

Sorvall Legend Micro 17 R microcentrifuge.

Isopropyl alcohol, for molecular biology >99 %: Sigma #I5916.

Ethyl alcohol, 200 proof for molecular biology: Sigma #E7023.

RNeasy Mini kit: Qiagen #74104.

RNase-free DNase set: Qiagen #79254.

NanoDrop 2000c spectrophotometer.

3   Methods

3.1   Tendon Decellularization

3.1.1   Phase 1: Collection Day

1. Harvest forelimb FDST tendons from donor animals using sterile surgical technique (*see* **Note 1**). Transfer tendons from necropsy suite to laboratory dissection table in a sterile vessel.

2. Dissect away surrounding tissue, leaving clean FDST for subsequent use. If samples are desired for RNA, remove a portion and immediately flash-freeze it in liquid nitrogen.

3. Transfer FDST samples to sterile storage containers and place in −80 °C chest freezer. Label containers with accession number, date and initials. Tendons for scaffold production may be stored in this manner indefinitely prior to use. Ensure materials are prepared for the next phase (*see* **Note 2**).

3.1.2   Phase 2: Dermatome Day

4. For best yield, tendons must remain frozen. If processing multiple tendons, remove only one sample from the freezer at a time. Thawed tendon will clog the dermatome.

5. Secure both ends of the tendon using locking clamps and firmly hold under tension.

6. Pass electric dermatome over the length of the tendon (*see* **Note 3**). Tendon slices are traditionally cut into 400 μm-thick ribbons. This parameter is adjustable on the side of the dermatome. The first few layers are discarded, and it may be useful to increase thickness for these passes.

7. Collect tendon ribbons in glass dissection tray filled with PBS. As ribbons accumulate or a new sample is prepared, transfer to 50 mL conical tubes and freeze at −20 °C.

8. Over the next few days, thaw and refreeze 50 mL conical tubes intended for use as scaffold material. Mark tubes with number of freeze–thaw cycles, up to four cycles. Reference samples may be set aside prior to this point in order to avoid further processing. Prepare solutions.

3.1.3   Phase 3: Detergent Decellularization

9. Bring samples, surgical instruments and dissection trays into biosafety cabinet. Trim tendon ribbons into pieces slightly larger than the provided template. Samples will expand during decellularization, but this ensures that final size will be sufficient.

10. Tendon pieces are individually decellularized in 6-well plates. Place pieces in wells and fill each well with 2 mL of 2 % SDS in 1 M Tris–HCl, pH 7.8. Label all plates.

11. Seal 6-well plates with paraffin film and transfer to refrigerated (4 °C) orbital shaker, set to 120 RPM. Leave for 24 h.

12. The next day, flip over all samples. Reseal plates and put back on shaker for 24 h.

13. After 24 h, add a small quantity of alcohol to the vacuum trap to prevent aspirated solution from bubbling over into main line. Aspirate SDS solution out of all wells.

14. Fill each well with 2 mL PBS. Place on shaker. After 20 min, aspirate solution. Repeat 5 times. Periodically flip samples over to ensure all areas are covered.

15. Add 1 mL trypsin–EDTA to each well. Place on refrigerated shaker for 5 min. Then transfer to incubator (37 °C) for another 5 min. Aspirate.

16. Wash each well with PBS twice as previously described.

17. Add sufficient volume of 0.05 mg/mL DNase I solution in PBS to cover all samples. Place on refrigerated shaker for 5 min, followed by the incubator for 10 min. Aspirate.

18. Wash each well with water twice as previously described for PBS.

19. Add 1 μl/mL protease inhibitor cocktail in PBS to each well. Place on shaker overnight.

20. Aspirate protease inhibitor cocktail, wash once with water, and replace with ethanol. Place on shaker for 2 h.

21. Aspirate, and fill all wells with water. Repeat twice more. Freeze in PBS.

22. Cut to size using a template. Templates can be printed, laminated, sterilized in alcohol and placed under a transparent plastic dissection tray in a biosafety cabinet.

3.2 Bioreactor Culture

1. Culture bone marrow-derived MSCs in monolayer using standard cell culture techniques (*see* **Note 4**).

2. Seed 250,000 cells onto the surface of each sample in 333 μL of differentiation medium (*see* **Note 5**).

3. Allow time for cells to settle and adhere, typically overnight.

4. Transfer samples to bioreactor vessels and initiate experimental protocol. This protocol will vary depending on experimental design.

5. When protocol is finished, harvest samples from bioreactor and immediately flash freeze in liquid nitrogen (*see* **Note 6**).

3.3 Sample Processing

1. Prepare lysis buffer stock: in 500 mL total volume of H_2O—250 g GITC, 17.6 mL 0.75 M sodium citrate buffer (11.03 g sodium citrate in 50 mL H_2O, bring down to pH 7.0 with NaOH) and 26.4 mL 10 % sarkosyl (in H_2O) (*see* **Note 7**).

2. Prior to milling, prepare 3 mL of lysis buffer in a 50 mL conical tube. Add 80 μL of β-mercaptoethanol.

3. Fill cryomill with liquid nitrogen up to the fill line (*see* **Note 8**).

4. Place sample (which is fresh or retrieved from storage at −80 °C) in impactor tube (*see* **Note 9**) and assemble. Immerse loaded tube in liquid nitrogen until sample is fully frozen and liquid nitrogen stops bubbling.

5. Insert sample tube into the cryomill. Run cryomill using the following settings—cycles: 2, precool: 1 min, run time: 1 min, cool time: 1 min, rate: 12 CPS.

6. Remove impactor tube from cryomill and ensure integrity of the vessel. Sample should be visibly powdered. If larger pieces remain, repeat milling for an additional cycle.

7. Open impactor tube using opening tool, transfer contents to 50 mL conical tube containing lysis buffer. Mix well and keep on ice.

8. If necessary for low-quantity samples, warm impactor tube to room temperature and rinse with additional lysis buffer. If not warmed, lysis buffer will freeze on contact.

9. Add 487.5 μL H_2O and 162.5 μL Triton X-100. Mix well. Leave on ice for 10 min.

10. Spin at $11,000 \times g$ for 5 min at 4 °C in standard benchtop centrifuge.

11. Transfer supernatant to a fresh Teflon 50 mL ultracentrifuge tube.

12. Add 8 ml of sodium acetate buffer, mix well. Then add 13 mL of saturated phenol and 3 mL chloroform (*see* **Note 10**). Mix by inversion and vortex for 10 s. Leave on ice for 10 min.

13. Balance tubes and centrifuge at $21,000 \times g$ for 15 min at 4 °C in high-speed centrifuge.

14. Remove upper aqueous phase and transfer to new ultracentrifuge tube. Do not allow contamination with interphase. Conduct a second collection close to the interphase and spin in microcentrifuge at $16,000 \times g$ for 10 min to retrieve residual.

15. Repeat extraction on aqueous phase with 8 mL saturated phenol and 8 mL chloroform until the aqueous phase is completely transparent.

16. Add 13 mL of isopropyl alcohol, mix well, and incubate at −20 °C overnight.

17. Centrifuge at $21,000 \times g$ in high-speed centrifuge. Small pellet may be visible, or not.

18. Aspirate supernatant and store open on bench to allow alcohol to evaporate.

19. Dissolve RNA pellet in 1 mL of H_2O in tube. Add 1 ml of lysis buffer RLT (containing 10 μL/mL β-mercaptoethanol) and 1 mL of ethyl alcohol. Mix by inversion.

20. Apply mixture to RNeasy spin column in batches until all of the solution is run through.

21. Follow spin column protocol from manufacturer. Include 15 min DNase step.

22. Perform final elution in 52 μL of H_2O. Use 2 μL to quantify RNA concentration in duplicate 1 μL samples in spectrophotometer. Store at -80 °C or immediately reverse-transcribe into cDNA.

4 Notes

1. Wear scrubs, disposable laboratory gowns, and BSL-2 PPE while handling tissue samples to protect yourself and the sterility of your samples.

2. The dermatome and plastic dissection trays must be sterilized in ethylene oxide. All other tools are sterilized in a steam autoclave. Solutions are filter-sterilized.

3. Use of the dermatome requires 2–3 participants. Plan a time when a group of people can process several samples in a session. However, do not allow dermatome to overheat.

4. An example standard maintenance medium recipe for bone marrow MSC culture is: low-glucose (1 g/L) DMEM with 110 μg/mL sodium pyruvate, 10 % MSC FBS and 100U/mL sodium penicillin. Differentiation medium is composed of maintenance medium plus 35.7 μg/mL L-ascorbic acid (alternate formulations are available [23]). 100 μg/mL streptomycin sulfate is optional for maintenance medium but should be excluded from differentiation medium.

5. Separating cell seeding into multiple steps can help prevent runoff and loss of cells.

6. Wear appropriate PPE when handling liquid nitrogen

7. Follow RNase avoidance procedures.

8. Schedule liquid nitrogen delivery in advance of cryomilling.

9. Impactor tubes have a limited lifespan. Use new tubes when available to avoid cracking and subsequent sample loss.

10. Use pipettes and centrifuge vials that can withstand phenol–chloroform extraction.

References

1. Magnusson SP, Langberg H, Kjaer M (2010) The pathogenesis of tendinopathy: balancing the response to loading. Nat Rev Rheumatol 6:262–268

2. Silver FH, Freeman JW, Seehra GP (2003) Collagen self-assembly and the development of tendon mechanical properties. J Biomech 36:1529–1553

3. Lavagnino M, Gardner K, Sedlak AM, Arnoczky SP (2013) Tendon cell ciliary length as a biomarker of in situ cytoskeletal tensional homeostasis. Muscles Ligaments Tendons J 3:118–121

4. Bi Y, Ehirchiou D, Kilts TM, Inkson CA, Embree MC, Sonoyama W, Li L, Leet AI, Seo BM, Zhang L, Shi S, Young MF (2007) Identification of tendon stem/progenitor cells and the role of the extracellular matrix in their niche. Nat Med 13:1219–1227

5. Wang N, Tytell JD, Ingber DE (2009) Mechanotransduction at a distance: mechanically coupling the extracellular matrix with the nucleus. Nat Rev Mol Cell Biol 10:75–82

6. Juneja SC, Veillette C (2013) Defects in tendon, ligament, and enthesis in response to genetic alterations in key proteoglycans and glycoproteins: a review. Arthritis 2013:154812

7. Mauck RL, Nicoll SB, Seyhan SL, Ateshian GA, Hung CT (2003) Synergistic action of growth factors and dynamic loading for articular cartilage tissue engineering. Tissue Eng 9:597–611

8. Kjaer M (2004) Role of extracellular matrix in adaptation of tendon and skeletal muscle to mechanical loading. Physiol Rev 84:649–698

9. Batson EL, Paramour RJ, Smith TJ, Birch HL, Patterson-Kane JC, Goodship AE (2003) Are the material properties and matrix composition of equine flexor and extensor tendons determined by their functions? Equine Vet J 35:314–318

10. Badylak SF (2002) The extracellular matrix as a scaffold for tissue reconstruction. Semin Cell Dev Biol 13:377–383

11. Yang G, Rothrauff BB, Lin H, Gottardi R, Alexander PG, Tuan RS (2013) Enhancement of tenogenic differentiation of human adipose stem cells by tendon-derived extracellular matrix. Biomaterials 34:9295–9306

12. Engler AJ, Sen S, Sweeney HL, Discher DE (2006) Matrix elasticity directs stem cell lineage specification. Cell 126:677–689

13. Egerbacher M, Arnoczky SP, Caballero O, Lavagnino M, Gardner KL (2008) Loss of homeostatic tension induces apoptosis in tendon cells: an in vitro study. Clin Orthop Relat Res 466:1562–1568

14. Altman GH, Horan RL, Martin I, Farhadi J, Stark PR, Volloch V, Richmond JC, Vunjak-Novakovic G, Kaplan DL (2002) Cell differentiation by mechanical stress. FASEB J 16:270–272

15. Banes AJ, Gilbert J, Taylor D, Monbureau O (1985) A new vacuum-operated stress-providing instrument that applies static or variable duration cyclic tension or compression to cells in vitro. J Cell Sci 75:35–42

16. Stewart AA, Barrett JG, Byron CR, Yates AC, Durgam SS, Evans RB, Stewart MC (2009) Comparison of equine tendon-, muscle-, and bone marrow-derived cells cultured on tendon matrix. Am J Vet Res 70:750–757

17. Cardwell RD, Dahlgren LA, Goldstein AS (2012) Electrospun fibre diameter, not alignment, affects mesenchymal stem cell differentiation into the tendon/ligament lineage. J Tissue Eng Regen Med 8(12):937–45

18. Kluge JA, Leisk GG, Cardwell RD, Fernandes AP, House M, Ward A, Dorfmann AL, Kaplan DL (2011) Bioreactor system using noninvasive imaging and mechanical stretch for biomaterial screening. Ann Biomed Eng 39:1390–1402

19. Youngstrom DW, Barrett JG (2015). Engineering tendon: scaffolds, bioreactors and models of regeneration. Stem Cells Int 2016:3919030

20. Screen HRC (2009) Hierarchical approaches to understanding tendon mechanics. J Biomech Sci Eng 4:481–499

21. Youngstrom DW, Barrett JG, Jose RR, Kaplan DL (2013) Functional characterization of detergent-decellularized equine tendon extracellular matrix for tissue engineering applications. PLoS One 8:e64151

22. Youngstrom DW, Rajpar I, Kaplan DL, Barrett JG (2015) A bioreactor system for in vitro tendon differentiation and tendon tissue engineering. J Orthop Res 33:911–918

23. Youngstrom DW, LaDow JE, Barrett JG (2015) Tenogenesis of bone marrow-, adipose- and tendon-derived stem cells in a dynamic bioreactor. Connect Tissue Res 2016 Mar 30:1–12 [Epub ahead of print]

Methods in Molecular Biology (2016) 1502: 203–211
DOI 10.1007/7651_2016_353
© Springer Science+Business Media New York 2016
Published online: 27 April 2016

Bioengineered Models of Solid Human Tumors for Cancer Research

Alessandro Marturano-Kruik*, Aranzazu Villasante*, and Gordana Vunjak-Novakovic

Abstract

The lack of controllable in vitro models that can recapitulate the features of solid tumors such as Ewing's sarcoma limits our understanding of the tumor initiation and progression and impedes the development of new therapies. Cancer research still relies of the use of simple cell culture, tumor spheroids, and small animals. Tissue-engineered tumor models are now being grown in vitro to mimic the actual tumors in patients. Recently, we have established a new protocol for bioengineering the Ewing's sarcoma, by infusing tumor cell aggregates into the human bone engineered from the patient's mesenchymal stem cells. The bone niche allows crosstalk between the tumor cells, osteoblasts and supporting cells of the bone, extracellular matrix, and the tissue microenvironment. The bioreactor platform used in these experiments also allows the implementation of physiologically relevant mechanical signals. Here, we describe a method to build an in vitro model of Ewing's sarcoma that mimics the key properties of the native tumor and provides the tissue context and physical regulatory signals.

Keywords: Human sarcoma, Tumor model, Cancer research, Bioreactor platform, Ewing's sarcoma

1 Introduction

Predictive models of human tumors are of paramount importance for studies of tumor initiation, progression and remission, identification of therapeutic targets, and development of new therapeutic modalities. Animal models have greatly contributed to our understanding of cancer, but their value in anticipating the effectiveness of treatment strategies in clinical trials remains uncertain [1]. In vitro testing of anticancer drugs typically involves growing cancer cell lines in monolayers on tissue-culture plastic dishes. Despite major progresses, monolayer cultures remain poor predictors of whether a given drug will be safe and ultimately provide clinical benefit [2, 3].

Most in vitro models, including tumor spheroids, cancer cells in scaffolds, and small cancer organoids lack the complexity of the native tumor milieu that mediates cancer processes [4–6].

---

* Author contributed equally with all other contributors.
The original version of this chapter was revised. An erratum to this chapter can be found at DOI 10.1007/978-1-4939-6478-9_5001.

Specifically, in the bone microenvironment, osteoblasts, osteoclasts, and mesenchymal stem cells (hMSC) mediate primary tumor growth and metastasis [7]. In addition, mechanical forces generated either from external sources or by cell contractions play important roles in cancer invasion into the bone [8, 9].

Bioengineering methods that have advanced stem cell research and regenerative medicine are becoming essential tools for cancer research [10]. Importantly, bioengineered tumors can provide cancer cells with a tissue context incorporating the extracellular matrix (ECM), supporting cells and physical signals.

In our effort to introduce substantial improvements over existing 3D models to study bone tumors we have developed a method to engineer in vitro human bone tumors, and to maintain these tumors in culture for prolonged periods of time (weeks to months). We cultured the Ewing's sarcoma (ES) cell spheroids within tissue engineered human bone. The bone was grown from adult hMSC capable of osteogenic differentiation within native bone ECM serving as a structural scaffold [11]. Additionally we exposed the cancer cells to biophysical *stimuli* mimicking those normally present within the bone tumor microenvironment, by using a bioreactor capable of providing mechanical compression [12].

We propose a model that allows cross talk between cancer cells and the key components of the bone tumor environment, such as mechanical forces and native mineralized ECM (Fig. 1).

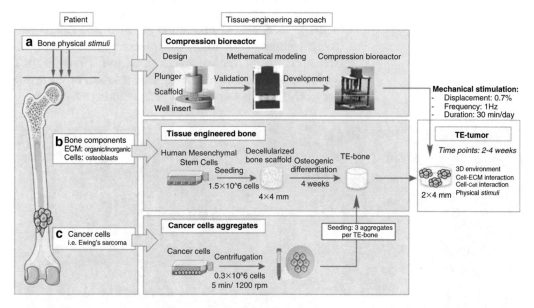

Fig. 1 Overall approach to the bioengineered bone tumor in vitro. The model recapitulates (**a**) physical stimuli in the patient's bone; (**b**) patient's bone comprising both organic and inorganic extracellular matrix (ECM) and bone cells (osteoblasts); (**c**) bone tumor compartment with Ewing's sarcoma cells. In order to mimic the tumor microenvironment and to generate a tissue-engineered tumor (TE-tumor), three cancer cell aggregates were infused into the tissue-engineered bone made of human mesenchymal stem cell derived osteoblasts. Then, the TE-tumor is cultured in a compression bioreactor and stimulated 30 min per day under physiological conditions (displacement 0.7 %; frequency 1 Hz) for 2 and 4 weeks

This bioengineered human tumor could dramatically improve upon the current preclinical drug-screening paradigm by providing valuable information about new therapeutic targets and anticancer drug efficacy.

2   Materials

2.1   Decellularized Bone Scaffold (Fig. 2b)

Prepare all solutions using milli-Q ultrapure water.

1. Driller (Dewat, cat. no DC970K2).

2. Diamond drill bites (Starlite Industries, cat. no 102045).

3. Fresh young bovine bone samples (Epiphysis of bovine femur) from a local slaughterhouse (Green Village Packing Co., New Jersey, USA).

4. $1\times$ PBS, 0.1 % (w/v) EDTA

5. Decellularization solutions: Prepare 1 L of a 10 mM Tris–HCl (pH 8) as stock solution (*see* **Note 1**).

   – *Solution 1:* 10 mM Tris–HCl (pH 8), 0.1 % (w/v) EDTA.

   – *Solution 2:* 10 mM Tris–HCl (pH 8), 0.5 % (w/v) SDS.

   – *DNAse/RNAse solution.* Using a syringe, take 5 mL of 10 mM Tris–HCl (pH 8) stock solution and inject it into a vial of DNAse (Life Technologies, cat. no 18068-015). Mix the solution by gentle vortex. Add the DNase solution (5 mL) to 35 mL of mQ water in a 50 mL Falcon tube (total volume = 40 mL), and incorporate 1 unit/mL RNAse A (Roche Applied Sciences, cat. no. 10109142001) (*see* **Note 2**). Mix the DNAse/RNAse solution by vortexing. Keep DNAse/RNAse solution at 4 °C until use.

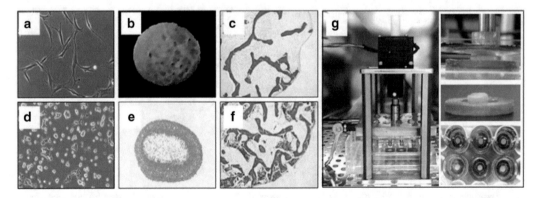

Fig. 2 Overview of key methods. (**a**) Human mesenchymal stem cells expanded in culture. (**b**) Decellularized bone scaffold used to seed mesenchymal cells and grow bone. (**c**) Scaffold seeded with mesenchymal stem cells to engineer bone (H&E stain). (**d**) Ewing's sarcoma cells (RD-ES line), expanded in culture. (**e**) Cancer cell aggregate. (**f**) Human Ewing's sarcoma tumor grown in vitro. (**g**) Apparatus for mechanical stimulation of engineered tumors

6. Caliber.

7. Dremel polish rotary tool.

2.2 Cells

Purchase Ewing's sarcoma cell lines from ATCC: RD-ES cell line (ATCC® Number: HTB-166) and SK-N-MC (ATCC® Number: HTB-10); purchase human mesenchymal stem cells (hMSC) from Lonza (Catalog number: PT-2501) (Fig. 2a, d).

2.3 Culture Media

Cancer media. Culture the RD-ES cells in ATCC-formulated RPMI-1640 Medium (RPMI) and the SK-N-MC cells in ATCC-formulated Eagle's Minimum Essential Medium (EMEM). Supplement both media with 10 % (v/v) Hyclone FBS, and 1 % penicillin–streptomycin.

hMSC medium. Basic medium for culturing hMSC consist in DMEM supplemented with 1 ng/mL Fibroblast Growth Factor (FGF), 10 % (v/v) Hyclone FBS, and 1 % penicillin–streptomycin.

Osteogenic medium. Prepare hMSC basic medium supplemented with 1 μM dexamethasone, 10 mM β-glycerophosphate, and 50 μM ascorbic acid-2-phosphate.

2.4 PDMS Rings

1. Polydimethylsiloxane (PDMS) and curing agent (KRAYDEN; cat. no DC2065622).

2. Glass petri dish.

3. Vacuum.

4. Oven at 60 °C.

5. 8 and 4 mm biopsy punch (VWR; cat. no 82030-348, cat. no 82030-354).

2.5 TE-Tumor

1. Non-treated 6-well plates (Nunc cat. no 150239).

2. 100 mm culture dish (Corning; cat. no 353003).

3. Sterile tweezers.

4. Sterile glass slides (Fisher; cat. no 12-548-5B).

5. Sterile razor blade (BD Bard-Parker, cat. no 371110).

2.6 Bioreactor: Design and Parts

The culture module:

1. 24-well plate (NUNC, cat. no 144530).

2. Well-insert: using a CNC milling machine mill a 30 × 30 × 3 mm (width × length × thickness) sheet of Ultem (McMaster Carr, cat. no 8685K47) to obtain a disk with a diameter of 15.5 mm and a central hole with a 6 mm in diameter and 1 mm deep. This insert is placed on the bottom of the culture well and allows the exact placement of the scaffold (*see* **Note 3**). Make at least 24.

3. The plunger: machine a polycarbonate rod (McMaster Car, cat. no 8571K31) to obtain a cylinder that is 8 mm in diameter and 30 mm in height. Make at least 24.

4. Lid: machine a polycarbonate bar (McMaster Car, cat. no 8574K321) to obtain a rectangular sheet of 127.89 × 85.60 mm (length × width) dimensions. Drill an 8 mm in diameter and 5 mm deep hole aligned to the center of the culture well. Repeat this for the remaining 23 wells.

5. Press fit the plungers into the drilled hole. To make sure all the plungers have the same height, face them down 0.5 mm using the milling machine.

The displacement module:

6. Using a shaft clamp (McMaster Car, cat. no 9660T3), connect the lid to a linear actuator controlled by a stepper motor (HaydonKerk 5700 series). The compression load is the result of the vertical motion provided by the linear actuator and the contact between the plungers and the scaffolds.

The control module and user interface:

7. The stepper motor is controlled by custom made microcontroller that includes an Arduino Mini Pro (Adafruit, cat. no 1501) and an A4988 stepper motor driver (Pololu, cat. no 1182) which are connected via a USB cable to a PC. The user can control all the experimental parameters form a custom made GUI (general user interface) created using open- source Arduino libraries.

2.7 Bioreactor: Mathematical Modeling

To characterize and predict forces generated in the TE-tumor upon exposure to mechanical loading in the bioreactor, run a finite element analysis in COMSOL® Multiphysics 4.2a. This analysis is can identify the stress field generated by a physiological-like loading on the TE-tumor. First, model the 3D geometry of the culture chamber, including the plunger, the well insert and a cylindrical scaffold (4 mm in diameter × 2 mm thick) using the drawing tool in the software. Second, insert the parameters related to the bone's mechanical properties such as: Young modulus (50 MPa), density (434 kg/m^3), and Poisson ratio (0.3). Run a quasi-static analysis that solves a time-dependent problem, assuming the structural mechanics component being static. Finally, solve the von Mises' tensor to evaluate the stress field generated in culture chamber.

3   Methods

3.1   TE-Bone

Culture hMSC in basic medium (DMEM supplemented with 10 % v/v Hyclone FBS and 1 % penicillin–streptomycin) for maintenance and expansion. Sterilize scaffolds in ethanol 70 % overnight. Suspend 1.5×10^6 hMSCs (passage 3) into 50 μL of medium and pipette the cells suspension onto the top of a blot-dried scaffold and allow to percolate through. After 15 min, rotate scaffold 180°, and add 10 μL of medium to prevent the cells from drying out. Repeat this process every 15 min for up to 2 h to facilitate uniform cell distribution. Culture the scaffolds in 6 mL of osteogenic medium for 4 weeks in a non-treated 6-well plate.

3.2   Cancer Cell Aggregates (Fig. 2e)

Centrifuge 0.3×10^6 Ewing's sarcoma cells in 4 mL of cancer medium in a 15 mL Falcon tube, at 290 g for 5 min. After centrifugation, culture the aggregates for 1 week at 37 °C in a humidified incubator, 5 % CO_2 (*see* **Note 4**).

3.3   PDMS Rings

1. Measure out 27 g of PDMS into a crystal petri dish.
2. Add 3 g of curing agent (ratio 9:1).
3. Gently mix them up for a few minutes with a spatula.
4. Carefully place the petri dish in the vacuum for 30 min until all the bubbles disappear from the solution.
5. Once all the bubbles have cleared from the mixture, carefully place the PDMS mold into the oven (*see* **Note 5**).
6. Let the samples bake for 1 h at 60 °C (or 24 h at room temperature).
7. Peel the PDMS off the mold using a sharp-tipped knife.
8. Punch the PDMS using a 8 mm biopsy punch for the inner part of the ring and a 4 mm biopsy punch for the core.
9. Autoclave the rings before using.

3.4   TE-Tumor (Fig. 2c, f)

1. Add 6 mL of cancer medium to 6-well plates (Nunc, cat. no150239). Keep them at 37 °C in a humidified incubator containing 5 % CO_2 until use.
2. Using sterile tweezers, place one glass slide on a 100 mm dish (*see* **Note 6**) and one PDMS ring onto the glass slide. Be sure that the PDMS ring is attached at the slide though the sticky side of the ring.
3. Harvest the cancer aggregates from the 15 mL Falcon tubes using a 5 mL pipette. Aspirate about 1 mL of medium carrying an aggregate and place it on a 100 mm culture dish. Repeat this step until harvesting all the aggregates.

4. Using sterile tweezers, place a TE-bone in an empty 100 mm dish. Bisect it (axial cut) using a razor blade. Immediately, bring one half to a well of 6-well plate with medium and culture it in the incubator.

5. Introduce one half of the TE-bone in the PDMS ring attached to the slide. Do not leave space between the bottom of the TE-bone and the crystal slide.

6. Using a 100 μL micropipette, load about 30–50 μL of medium containing three cancer aggregates onto the TE-Bone placed into the PDMS ring.

7. Using a tweezers, flip out the ring with the TE-Bone-cancer aggregates. Push gently the TE-bone against the slide and let the aggregates get into the construct. Again, do not leave space between the bottom of the TE-bone and the crystal slide.

8. Discard the ring and place the TE-tumor in a well of 6-well plate with medium at 37 °C in a humidified incubator containing 5 % CO_2 during 2 weeks.

9. Change medium biweekly.

3.5 Bioreactor Stimulation (Fig. 2g)

1. Using sterile tweezers place the well inserts at the bottom of a 24-well plate (Nunc, cat. no 144530). Add 2 mL of cancer medium to each well. Keep the plate at 37 °C in a humidified incubator containing 5 % CO_2 until use.

2. Using sterile tweezers transfer the TE-tumor in the 24 well plate. Place the scaffold in the center of the well insert (*see* **Note 7**).

3. Place the 24 well plate in the center of the bioreactor.

4. Lower the lid and plungers using the controls on the GUI. Make sure the lid covers completely the plate.

5. Place the bioreactor at 37 °C in a humidified incubator containing 5 % CO_2.

6. The compression protocol consisted in 24 h of culture in the bioreactor with three loading inputs. The first application consisted of 0.7 % of strain (for a 2 mm thick scaffold it equals 14 μm of displacement amplitude), applied using a sinusoidal wave form for the vertical motion at 1 Hz frequency for 1800 loading cycles (equivalent of 30 min of stimulation) (*see* **Note 8**).

7. Stimulate sample right after the relocation into the bioreactor, let rest overnight and stimulate again, both in the morning and in the afternoon (*see* **Note 9**).

4 Notes

1. Add 10 mL of a 1 M Tris–HCl, pH 8, stock solution, (Thermo Fisher scientific; 15568-025), to a 1-L graduated cylinder. Mix with 990 mL of mQ water and adjust pH using HCl. Store at room temperature.

2. Close the screw-top caps loosely (Do NOT close all the way so that gas can be exchanged).

3. Make sure the scaffold is well centered. Use the well insert off-set feature as a reference.

4. A flat paper or aluminum foil layer placed underneath the petri dish is a good idea just in case the PDMS overflows and sticks to the oven surface.

5. This protocol uses a 6-well plate format. Amounts may be scaled up or down if using another size format.

6. You can also use the lid of a 100 mm culture dish.

7. Make sure the well insert is placed correctly at the bottom of the well.

8. Various stimulation protocols can be used. It is possible to tune several culture parameters such as frequency, strain and time depending on the user's need.

9. Tune stimulation cycles according to experimental requirements.

Acknowledgement

We gratefully acknowledge funding support by National Institutes of Health (grants UH3EB017103 and EB002520) and the Alfonso Martin Escudero Foundation.

References

1. van der Worp HB, Howells DW, Sena ES, Porritt MJ, Rewell S, O'Collins V, Macleod MR (2010) Can animal models of disease reliably inform human studies? PLoS Med 7(3): e1000245. doi:10.1371/journal.pmed. 1000245

2. Voskoglou-Nomikos T, Pater JL, Seymour L (2003) Clinical predictive value of the in vitro cell line, human xenograft, and mouse allograft preclinical cancer models. Clin Cancer Res 9 (11):4227–4239

3. Hutchinson L, Kirk R (2011) High drug attrition rates—where are we going wrong? Nat Rev Clin Oncol 8(4):189–190. doi:10.1038/ nrclinonc.2011.34

4. Correia AL, Bissell MJ (2012) The tumor microenvironment is a dominant force in multidrug resistance. Drug Resist Updat 15 (1–2):39–49. doi:10.1016/j.drup.2012.01. 006

5. Bissell MJ, Hines WC (2011) Why don't we get more cancer? A proposed role of the microenvironment in restraining cancer progression. Nat Med 17(3):320–329. doi:10.1038/nm. 2328

6. Hanahan D, Coussens LM (2012) Accessories to the crime: functions of cells recruited to the tumor microenvironment. Cancer Cell 21 (3):309–322. doi:10.1016/j.ccr.2012.02.022

7. Villasante A, Vunjak-Novakovic G (2015) Bioengineered tumors. Bioengineered 6 (2):73–76. doi:10.1080/21655979.2015. 1011039

8. Lynch ME, Brooks D, Mohanan S, Lee MJ, Polamraju P, Dent K, Bonassar LJ, van der Meulen MC, Fischbach C (2013) In vivo tibial compression decreases osteolysis and tumor formation in a human metastatic breast cancer model. J Bone Miner Res 28(11):2357–2367. doi:10.1002/jbmr.1966

9. Tse JM, Cheng G, Tyrrell JA, Wilcox-Adelman SA, Boucher Y, Jain RK, Munn LL (2012) Mechanical compression drives cancer cells toward invasive phenotype. Proc Natl Acad Sci U S A 109(3):911–916. doi:10.1073/ pnas.1118910109

10. Villasante A, Vunjak-Novakovic G (2015) Tissue-engineered models of human tumors for cancer research. Expert Opin Drug Discov 10(3):257–268. doi:10.1517/17460441. 2015.1009442

11. Villasante A, Marturano-Kruik A, Vunjak-Novakovic G (2014) Bioengineered human tumor within a bone niche. Biomaterials 35 (22):5785–5794. doi:10.1016/j.biomaterials. 2014.03.081

12. Marturano-Kruik A, Yeager K, Bach D, Villasante A, Cimetta E, Vunjak-Novakovic G (2015) Mimicking biophysical stimuli within bone tumor microenvironment. Conf Proc IEEE Eng Med Biol Soc 2015:3561–3564. doi:10.1109/EMBC.2015.7319162

Methods in Molecular Biology (2016) 1502: 213–221
DOI 10.1007/7651_2015_309
© Springer Science+Business Media New York 2015
Published online: 13 December 2015

Development of a Bladder Bioreactor for Tissue Engineering in Urology

Niall F. Davis and Anthony Callanan

Abstract

A urinary bladder bioreactor was constructed to replicate physiological bladder dynamics. A cyclical low-delivery pressure regulator mimicked filling pressures of the human bladder. Cell growth was evaluated by culturing human urothelial cells (UCs) on porcine extracellular matrix scaffolds (ECMs) in the bioreactor and in static growth conditions for 5 consecutive days. UC proliferation was compared with quantitative viability indicators and by fluorescent markers for intracellular esterase activity and plasma membrane integrity. Scaffold integrity was characterized with scanning electron microscopy and 4,6-diamidino-2-phenylindole staining.

Keywords: Bladder bioreactor, Tissue-engineering, Regenerative medicine, Urothelial cells

1 Introduction

Surgical repair for end-stage bladder disease often utilizes vascularized, autogenous, mucus-secreting gastrointestinal tissue to either replace the diseased organ or augment inadequate bladder tissue [1]. Augmentation cystoplasty is a urological procedure that involves the addition of viscoelastic ileal tissue to defective bladder tissue to improve functional bladder capacity [2]. Postoperatively, the compliant smooth muscle of the bowel is often sufficient to restore the basic shape, structure, and function of the urinary bladder: however lifelong postoperative complications are common. Morbidities that result from interposition of intestinal tissue may be divided into three broad areas: metabolic, neuromechanical, and technical-surgical [3]. Metabolic complications are the result of altered solute reabsorption by the intestine of the urine that it contains. Neuromechanical aspects involve the configuration of the bowel which affects storage volume and contraction of the intestine that may lead to difficulties in storage. Finally, technical-surgical complications involve aspects of the procedure that result in surgical morbidity.

ECMs are decellularized biological scaffolds commonly derived from porcine organs [4]. After their preparation process decellularized ECM scaffolds may be seeded with autologous impermeable urothelial cells (UCs) in vitro [5]. Cell-seeded scaffolds offer important advantages in urological settings when compared to gastrointestinal segments as UCs may provide a protective impermeable epithelial lining prior to in vivo implantation [3]. Limitations associated with cell-seeded scaffolds are harvesting difficulties and low proliferative potential in vitro [6, 7]. Theoretically "bladder bioreactors" may improve urothelial cell proliferation by preconditioning cell-seeded scaffolds in a simulated in vivo physiological environment [8]. The objective of the present study is to construct and evaluate a urinary bladder bioreactor for urological tissue engineering purposes.

2   Materials

2.1   Construction of Bioreactor

Components were purchased from BOC gases® (UK and Ireland) unless otherwise indicated. Primary constituents included a sealed pressure chamber with a transparent window, pressurized gas containers, and silicone tubing (Centre for Applied Biomedical Engineering and Research [CABER], University of Limerick, Ireland) (Fig. 1). The principal aim of the bioreactor was to increase hydrostatic pressure on cell-seeded ECM scaffolds from 0 to 10 cm H_2O via a cyclical low-delivery pressure regulator over preset durations [9]. Urothelial cell viability and proliferative activity on ECM scaffolds were measured and compared for 5 consecutive days in a control and bioreactor experimental group by quantitative viability indicators [8].

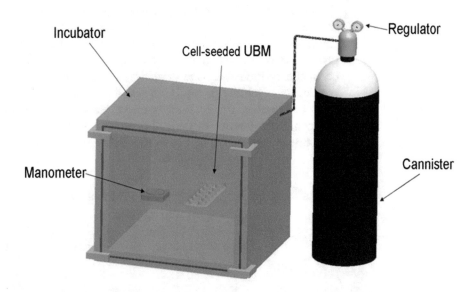

Fig. 1 Simplified schematic of bladder bioreactor

2.2 Culture and Expansion of UCs

Cell culture materials were obtained from Innovative Technologies in Biological Systems, Bizkaia, Spain (Innoprot®) unless indicated [10].

1. UCs were initially cultured under standard cell culture conditions in a humidified atmosphere of 95 % air and 5 % CO_2 at 37 °C in T-75 vented cap flasks (Sarstedt Ltd, Wexford, Ireland).

2. The cell line was grown to confluence in urothelial cell medium (UCM) and supplemented with 1 % urothelial cell growth supplement (UCGS) and 1 % penicillin/streptomycin solution (P/S solution). For expansion, cells were fed every 48 h until 85–90 % confluency, which was usually achieved after 7–8 days.

3. Confluent cells were harvested by incubation with 0.0025 % trypsin/0.5 mM EDTA Solution (T/E solution). Trypsin was neutralized with Trypsin Neutralization Solution (TNS) that contained buffered saline solution and 10 % fetal bovine serum (FBS) and was calcium and magnesium free. The cells were then centrifuged for 5 min (1000 g force) and gathered. Passage 4 cells were used in the study.

2.3 Measurement of Urothelial Cell Viability

Cell viability was assessed and quantified with the alamarBlue® cell viability reagent (Invitrogen, Ireland) for 5 consecutive days. In the presence of viable cells the dye reduces, turns red, and becomes highly fluorescent.

1. One-tenth of the volume of alamarBlue® reagent was added directly into culture medium based on the protocol demonstrated in Table 1.

2. Proliferation rates in both experimental groups were measured as a fold increase in the number of viable cells per day. To eliminate the potential for intervention bias between both groups, two control groups were created. Control group 1 consisted of increasing cell densities cultured in static growth conditions. Control group 2 consisted of increasing cell densities cultured in the bioreactor under static conditions (i.e., without the influence of cyclical pressure). Calibration curves with linear fit were applied to both control groups (Fig. 2).

Table 1
AlamarBlue® protocol for assessing viable cell numbers

Format	Volume of cells and medium	Volume of 10× alamarBlue® to add
Cuvette	1 mL	100 µL
96-well plate	100 µL	10 µL
284-well plate	40 µL	4 µL

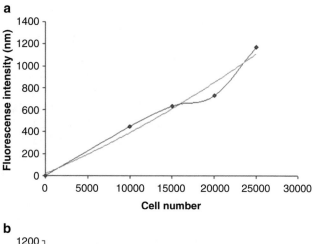

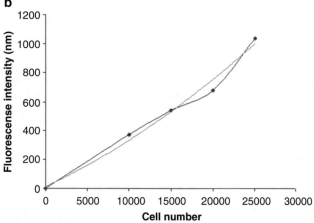

Fig. 2 Applying standard curves eliminated intervention bias between both experimental groups. (**a**) Calibration curve for human urothelial cells cultured under static growth conditions. (**b**) Calibration curve for human urothelial cells cultured in the bioreactor under static growth conditions

2.4 Scanning Electron Microscopy

To assess the structural integrity of UBM scaffolds in the bioreactor scanning electron microscopy (SEM) was performed before and after 5 days on UBM with and without cells. The samples were sputter coated with a thin layer of gold and palladium.

2.5 4-6-Diamidino-2-phenylindole (DAPI) Staining

4-6-Diamidino-2-phenylindole (DAPI) nucleic acid staining (Invitrogen®, Dublin, Ireland) was performed to confirm the presence of cell nuclei in the bioreactor group after 5 days of growth (*see* **Note 1**).

2.6 Measurement of Fluorescence Activity

"Viability/cytotoxicity" fluorescence assays were performed by means of three-dimensional confocal microscopy on both experimental groups to compare cell viability after 5 days of culture (LSM 700, Carl Zeiss Microimaging, Gottingen, Germany). Live cells were distinguished by the presence of ubiquitous intracellular esterase activity, determined by the enzymatic conversion of the virtually nonfluorescent cell-permeant calcein-AM to the intensely fluorescent calcein (*see* **Note 2**).

EthD-1 enters cells with damaged membranes and undergoes a 40-fold enhancement of fluorescence upon binding to nucleic acids, thereby producing a bright red fluorescence in dead cells (ex/em ~495 nm/~635 nm). EthD-1 is excluded by the intact plasma membrane of live cells. The determination of cell viability depends on these physical and biochemical properties of cells. Cytotoxic events that do not affect these cell properties may not be accurately assessed using this method. Background fluorescence levels are inherently low with this assay technique because the dyes are virtually nonfluorescent before interacting with cells.

3 Methods

3.1 Bioreactor and Physiological Intravesical Pressures

1. A manometer was placed in the sealed pressure chamber for the duration of the study to accurately monitor pressures within the bioreactor throughout the entire experimental time period.

2. Cyclical filling pressures of the human bladder were mimicked by connecting a low delivery pressure regulator, with a pressure-adjusting screw, to the gas container (95 % air and 5 % CO_2) that increased the fluid pressure on the cell-seeded UBM scaffold within the bioreactor from 0 to 10 cm H_2O (Fig. 3).

3. Application of cyclical pressure over a preset duration resulted in mechanical strain on the cell-seeded ECM scaffold. Voiding pressures were mimicked by releasing the infused gas from a "Y-connector" silicon outlet source over a period of approximately 10 s every 6–8 h [8].

3.2 Urothelial Cell-Seeding Techniques in Control and Experimental Group

1. Specimens of urinary bladder matrix (UBM), a porcine extracellular matrix graft, were cut into circles of 2 cm diameter, transferred into 12-well tissue culture plates and weighted with stainless steel rings also 2 cm in diameter to inhibit their lifting. Preparation of UBM has previously been described (*see* **Note 3**) [11]. Each stainless steel ring was autoclaved prior to insertion to ensure sterility. UBM scaffolds were then seeded with urothelial cells ($2.5–5.0 \times 10^4$ cells/cm^2 per well).

2. Equally matched circles of UBM were inserted into a second modified 6-well plate and transferred into the urinary bladder bioreactor for comparative assessment. Luminal surfaces of UBM scaffolds were seeded with 5.0×10^4 cells/cm^2 per well and cultured in 2 ml of UCM

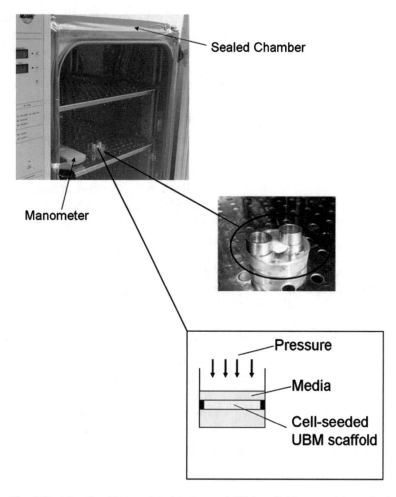

Fig. 3 Photograph of bioreactor chamber and ECM scaffold exposed to cyclical physiological urinary bladder pressures

3.3 Measurement of Urothelial Cell Viability

1. After adding the alamarBlue® reagent, the solution was incubated for 4 h at 37 °C in a cell culture incubator that is protected from direct light (*see* **Note 4**).

2. For greater sensitivity the extent of alamarBlue® dye reduction was quantified by fluorescence spectrophotometry (BioTek, Synergy HT Multi-Mode Microplate Reader). Proliferation rate was measured as a fold increase in the number of viable cells per day.

3. Fluorescence was read using a fluorescence excitation wavelength of 540–570 nm (peak excitation is 570 nm). Fluorescence emission was read at 580–610 nm (peak emission was 585 nm). Fluorescence methods are more sensitive than absorbance methods and this method was applied to all experiments (*see* **Note 5**).

4. Absorbance of alamarBlue® can also be monitored at 570 nm, using 600 nm as a reference wavelength value. Average 600 nm absorbance values of the cell culture medium alone (background) were subtracted from the 570 nm absorbance values of experimental wells. Background-subtracted 570 nm absorbance versus concentration of the test compound values were then plotted.

5. Cells were plated in 100 μL medium into 96-well tissue culture plates after performing cell number titration in the range of 40–10,000 for adherent cells and 2000–500,000 for suspension cells. A background control of 100 μL of medium without cells should be used. Subsequently 10 μL of alamarBlue® was added into the medium and the cells incubated at 37 °C overnight.

3.4 Scanning Electron Microscopy

1. Initially, primary fixation was performed by immersing the sample in 2.5 % glutaraldehyde in 0.1 M cacodylate buffer, pH 7.4 for 2 h at room temperature or in a refrigerator at 4 °C overnight.

2. Samples were washed for 5 min on three consecutive occasions in 0.1 M cacodylate buffer (pH 7.4).

3. Secondary fixation was performed by immersing the sample in 1 % osmium tetroxide (aqueous) pH 7.4 for 1 h at room temperature in a lightproof container. The washing process was repeated in 0.1 M cacodylate buffer pH 7.4.

4. Dehydration was achieved by the following method (*see* **Note 6**):
   (a) 1 × 10 min in 25 % ethanol
   (b) 1 × 10 min in 50 % ethanol
   (c) 1 × 10 min in 70 % ethanol
   (d) 1 × 10 min in 85 % ethanol
   (e) 1 × 10 min in 95 % ethanol
   (f) 2 × 10 min in 100 % ethanol
   (g) 1 × 10 min 100 % ethanol (EM grade)

5. Sputter coating was performed by applying a thin layer of gold and palladium over the sample with an automated sputter coater. This process can take up to 10 min.

3.5 4-6-Diamidino-2-phenylindole Staining

1. To make a 5 mg/mL DAPI stock solution (14.3 mM for the dihydrochloride or 10.9 mM for the dilactate), dissolve the contents of one vial (10 mg) in 2 mL of deionized water (dH₂O) or dimethylformamide (DMF). The less water-soluble DAPI dihydrochloride may take some time to completely dissolve in water and sonication may be necessary.

2. Counterstaining protocol:

(a) Equilibrate the sample briefly with phosphate-buffered saline (PBS).

(b) Dilute the DAPI stock solution to 300 nM in PBS. Add approximately 300 μL of this dilute DAPI staining solution to the cover slip preparation, making certain that the cells are completely covered.

(c) Incubate for 1–5 min.

(d) Rinse the sample several times in PBS. Drain excess buffer from the cover slip and mount.

(e) View the sample using fluorescence microscopy with appropriate filters.

3.6 Measurement of Fluorescence Activity

1. Medium was aspirated and the culture vessel washed with DPBS. Five milliliters of DPBS were added into a 15 ml centrifuge tube followed by 10 μL of ethidium homodimer-1 (Ethd-1) and 10 μL of calcein to the tube (concentrations may change depending on the sample size).

2. The newly mixed solution was added to the cell sample and placed in a refrigerator at 4 °C for 10 min. After this time frame, the solution was aspirated and the sample washed with PBS. A small amount of PBS was left on the sample to prevent dehydration.

4 Notes

1. DAPI staining was performed by dissolving 10 mg of DAPI into 2 mL of deionized water and protecting the solution from light. Subsequently, 300 μL of the diluted DAPI staining solution was added to the cell-seeded scaffolds. The samples were then incubated for 5 min and evaluated with fluorescence confocal microscopy (LSM 700, Carl Zeiss Microimaging, Gottingen, Germany).

2. The polyanionic dye calcein is well retained within live cells, producing an intense uniform green fluorescence in live cells (ex/em ~495 nm/~515 nm) [12].

3. The harvested porcine bladder was rinsed in normal saline to remove the inner luminal urothelial cell lining. The outer abluminal muscular layers were removed by manual delamination. The specimen was bisected along one side to form a sheet. Subsequently, the tunica serosa, tunica muscularis externa, tunica submucosa, and muscularis mucosa were manually removed while preserving an intact basement membrane and lamina propria. The decellularization process was completed by

soaking the matrix in buffered saline (pH 7.4), placing it in peracetic acid/4 % ethanol for 2 h and rinsing it in sterile buffered saline. Finally, sterilization was achieved by exposure to ethylene oxide [13, 14].

4. Sensitivity of detection generally increases with longer duration times; therefore samples with fewer cells should use incubation times up to 24 h.

5. Assay plates can be wrapped in foil, stored at 4 °C, and read within 1–3 days without affecting the fluorescence or absorbance values.

6. It takes approximately 40 min to achieve a critical dry point and the sample can then be mounted onto a metal stub with a double-sided carbon tape.

References

1. Biers SM, Venn SN, Greenwell TJ (2012) The past, present and future of augmentation cystoplasty. BJU Int 109:1280–1293

2. Davis NF, Mooney R, Callanan A et al (2011) Augmentation cystoplasty and extracellular matrix scaffolds: an ex vivo comparative study with autogenous detubularised ileum. PLoS One 6(5):1–7

3. Flood HD, Malhotra SJ, O'Connell HE et al (1995) Long-term results and complications using augmentation cystoplasty in reconstructive urology. Neurourol Urodyn 14 (4):297–309

4. Crapo PM, Gilbert TW, Badylak SF (2011) An overview of tissue and whole organ decellularization processes. Biomaterials 32:3233–3243

5. Davis NF, McGuire BB, Callanan A et al (2010) Xenogenic extracellular matrices as potential biomaterials for interposition grafting in urological surgery. J Urol 184 (6):2246–2253

6. Zhang Y, McNeill E, Tian H et al (2008) Urine derived cells are a potential source for urological tissue reconstruction. J Urol 180 (5):2226–2233

7. Atala A, Bauer SB, Soker S et al (2006) Tissue-engineered autologous bladders for patients needing cystoplasty. Lancet 367:1241–1246

8. Davis NF, Mooney R, Piterina AV et al (2011) Construction and evaluation of urinary bladder bioreactor for urologic tissue-engineering purposes. Urology 78(4):954–960

9. Wallis MC, Yeger H, Cartwright L et al (2008) Feasibility study of a novel urinary bladder bioreactor. Tissue Eng Part A 14(3):339–348

10. Davis NF, Callanan A, McGuire BB et al (2011) Evaluation of viability and proliferative activity of human urothelial cells cultured onto xenogenic tissue-engineered extracellular matrices. Urology 77(4):1007.e1–1007.e7

11. Callanan A, Davis NF, Walsh MT et al (2012) Mechanical characterisation of unidirectional and cross-directional multilayered urinary bladder matrix (UBM) scaffolds. Med Eng Phys 34(9):1368–1374

12. Shaikh FM, O'Brien TP, Callanan A et al (2010) New pulsatile hydrostatic pressure bioreactor for vascular tissue-engineered constructs. Artif Organs 34(2):153–158

13. Brown B, Lindberg K, Reing J et al (2006) The basement membrane component of biologic scaffolds derived from extracellular matrix. Tissue Eng 12(3):519–526

14. Gilbert TW, Sellaro TL, Badylak SF (2006) Decellularization of tissues and organs. Biomaterials 27:3675–3683

Methods in Molecular Biology (2016) 1502: 223–235
DOI 10.1007/7651_2016_337
© Springer Science+Business Media New York 2016
Published online: 10 April 2016

Use of Microfluidic Technology to Monitor the Differentiation and Migration of Human ESC-Derived Neural Cells

Jiwoo Bae, Nayeon Lee, Wankyu Choi, Suji Lee, Jung Jae Ko, Baek Soo Han, Sang Chul Lee, Noo Li Jeon, and Jihwan Song

Abstract

Microfluidics forms the basis of unique experimental approaches that visualize the development of neural structure using micro-scale devices and aids the guidance of neurite growth in an axonal isolation compartment. We utilized microfluidics technology to monitor the differentiation and migration of neural cells derived from human embryonic stems cells (hESC). We cocultured hESC with PA6 stromal cells and isolated neural rosette-like structures, which subsequently formed neurospheres in a suspension culture. We found that Tuj1-positive neural cells but not nestin-positive neural precursor cells (NPC) were able to enter the microfluidics grooves (microchannels), suggesting a neural cell-migratory capacity that was dependent on neuronal differentiation. We also showed that bundles of axons formed and extended into the microchannels.

Taken together, these results demonstrated that microfluidics technology can provide useful tools to study neurite outgrowth and axon guidance of neural cells, which are derived from human embryonic stem cells.

Keywords: Human embryonic stem cells (hESCs), Neural differentiation, Neural precursor cells (NPCs), Mature neurons, Neurite outgrowth, Microfluidic devices

1 Introduction

Human embryonic stem cells (hESCs) derived from the inner cell mass of the pre-implantation-stage embryo have the potential to proliferate indefinitely and to differentiate into any cells that constitute the body [1, 2]. For these reasons, hESCs have been widely regarded as ideal sources for cell replacement therapies to treat various forms of degenerative, metabolic, and genetic diseases, including neurological disorders. However, despite extensive research on stem cells, there is still a lack of understanding on the regulation of cell proliferation and differentiation [3]. Components of the stem cell microenvironment include growth factors, extracellular matrix, temperature, and pH. Unlike the unpredictable microenvironment of cells, microfluidics allows the formation of a controlled microenvironment.

Microfluidics is a new technology platform for cellular neuroscience that deals with the flow of liquids in channels of micrometer size. These experimental tools offer advantages for visualizing the development of neural structures within a micro-scale device, which allow the guidance of neurite growth into an axonal isolation compartment. Also, microfluidic devices share physiologically relevant features of fluid flows generation, such as the maintenance of a constant soluble microenvironment [4]. In addition to this, these devices can be applied with gradient materials, which allows a potentially unlimited range of concentrations of multi-inputs into the cellular microenvironment [5].

Recently, a number of microfluidic technologies for studying neural development have been developed [5, 6]. Taylor's group devised a neuronal culture in microfluidic devices that enables the culture and isolation of cell bodies and axons [7, 8]. And another group has developed a classified microfluidic neuronal culture system for signaling studies [9].

Using the microfluidic technology, we cocultured hESCs with PA8 stromal cells, which became differentiated into mature neuron, and observed migration, as well as expanded neurite outgrowth of the differentiated neural cells (Fig. 1). These results suggest that

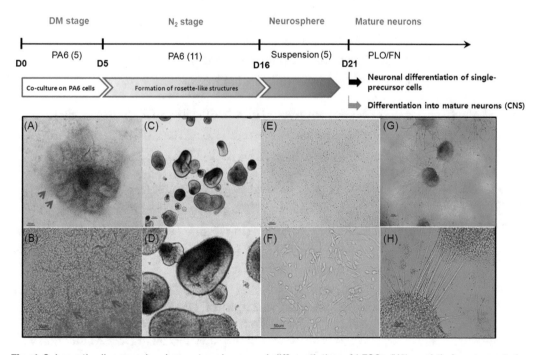

Fig. 1 Schematic diagram showing a stepwise neural differentiation of hESCs (H9), and their representative morphologies at each stage. *Upper panels* (**A**, **C**, **E**, and **G**) show low magnifications, with enlarged images in the *bottom panels* (**B**, **D**, **F**, and **H**), respectively. (**A**, **B**) Rosette-like structures formed by coculturing hESCs with PA6 stromal cells. (**C**, **D**) Neurospheres formed in suspension culture. (**E**, **F**) Dissociated neural precursor cells maintained on PLO/FN double-coated dishes. (**G**, **H**) Neurospheres differentiated into mature neurons in N2 medium containing BDNF. Abbreviations: hESCs, human embryonic stem cells; PLO/FN, poly-L-ornithine and fibronectin; BDNF, brain-derived neurotrophic factor (Reproduced with permission from ref. [11])

our system can provide critical guidelines for neural differentiation using microfluidic devices to study disease pathogenesis or to screen drugs when disease-specific cell lines are used.

2 Materials

2.1 Microfluidic Devices

1. PDMS with curing agent (Dow Hitech; Sylgard 184).
2. Poly-L-ornithine solution (PLO; 1 mg/ml, Sigma) in distilled water. Store at 4 °C.
3. Fibronectin solution (FN; 1 mg/ml, Sigma) in distilled water. Store at 4 °C.
4. DI water.
5. Digital balance.
6. Hot plate.
7. Vacuum desiccator.
8. Plasma cleaner (Harrick Scientific PDC-001).
9. Disposable plastic cups.
10. Scotch tape (3 M Scotch 471).
11. Glass coverslips (YMS Korea 24 × 40 mm).
12. Surgical blade.
13. Tissue biopsy punches (6 mm).
14. Silicon wafer (Silicon Inc.).

2.2 Cell Lines

1. Mouse embryonic fibroblast cells (MEF, primary cells).
2. PA6 stromal feeder cells (RCB1127: MC3T3-G2/PA6).
3. H9 human ES cells (WiCell, USA).

2.3 Preparation of MMC Treated-MEF Medium

1. Dulbecco's modified Eagle's medium (DMEM; Invitrogen).
2. Fetal bovine serum (FBS; Welgene) (see **Note 1**).
3. Penicillin–streptomycin solution (P/S; Welgene). P/S solution consists of 100 U/ml penicillin and 100 µg/ml streptomycin.
4. β-mercaptoethanol (Gibco).
5. Mitomycin C solution (MMC; 2 mg, Invitrogen). Dissolve MMC solution in 20 ml of DMEM medium and sterilize it using a 0.22 µm filter. Aliquots of 1 ml are stored at −20 °C. When ready to use, dilute to 10 µg/ml in DMEM medium with P/S (see **Note 2**).
6. Phosphate-buffered saline solution without calcium and magnesium (PBS; Invitrogen).
7. Trypsin–EDTA solution (0.25 %, Welgene, Korea).

8. Hemocytometer (Millipore).

9. 35 mm^2 culture dish (BD Biosciences).

10. 37 °C with 5 % CO_2 incubator (Sanyo, Japan).

2.4 Preparation of Human ESCs Medium

1. Dulbecco's modified Eagle's medium F-12 (DMEM F-12; Invitrogen).

2. Knockout SR serum replacement (KOSR; Invitrogen). Aliquots of 40 ml are stored at −20 °C. Thaw at −4 °C for use.

3. Nonessential amino acid solution (NEAA, 0.1 mM; Welgene).

4. Recombinant human FGF-basic (bFGF; Peprotech). The stock solution of 10 μg/ml is made in PBS (−) or DMEM F-12 medium and aliquots of 0.1 ml are stored at −20 °C. Working concentration is 4 ng/ml.

5. CTK solution including collagenase type IV (Gibco, USA), Trypsin (Gibco, USA), and Knockout SR (Gibco, USA).

6. Phosphate-buffered saline solution without calcium and magnesium (PBS; Invitrogen).

7. 35 mm^2 culture dish (BD biosciences).

8. 37 °C with 5 % CO_2 incubator.

2.5 Preparation of PA6 Stromal Cell Medium

1. Minimum essential medium, alpha modification (Alpha MEM; Welgene).

2. Fetal bovine serum (FBS; Welgene).

3. Penicillin–streptomycin solution (P/S; Welgene). P/S solution consists of 100 U/ml penicillin and 100 μg/ml streptomycin.

4. β-mercaptoethanol (Gibco).

5. Collagen type I solution (5 mg/ml, Gibco).

6. Acetic acid (1 N, Sigma).

7. Phosphate-buffered saline solution without calcium and magnesium (PBS; Invitrogen).

8. Trypsin–EDTA solution (0.25 %, Welgene).

9. Hemocytometer (Millipore).

10. 60 mm^2 culture dish (BD Biosciences).

11. 37 °C with 5 % CO_2 incubator.

12. *Pasteur* pipette.

2.6 Neural Differentiation Medium

The basal medium for neural differentiation is Glasgow's Modified Minimum Essential Medium (GMEM; Invitrogen) containing 1 % penicillin, 1 % streptomycin, 1 % nonessential amino acids, 0.1 % ß-mercaptoethanol with 3 mM D-glucose (Sigma), 0.2 mM ascorbic acid (Sigma), and 2 mM L-glutamine (Welgene).

1. Knockout SR serum replacement (KOSR; Invitrogen).

2. N2 supplement (100×; Invitrogen).

3. 10 ng/ml recombinant Human FGF-basic (bFGF; PeproTech).

4. 10 ng/ml recombinant human epidermal growth factor (EGF; PeproTech).

5. 20 ng/ml recombinant human brain-derived neurotropic factor (BDNF; R&D systems).

Neural medium	Serum	Replacement	Factor
DM medium	10 % KOSR	–	–
N2 medium	–	N2 supplement (100×; Invitrogen)	–
Neurosphere medium	–	N2 supplement (100×; Invitrogen)	10 ng/ml bFGF (Peprotech) 10 ng/ml EGF (Peprotech)
Mature medium	–	N2 supplement (100×; Invitrogen)	20 ng/ml BDNF (R&D systems)

2.7 Immunocytochemistry

1. Fixation solution: 4 % formaldehyde (FA; Sigma) in PBS with calcium and magnesium (PBS (+); Welgene).

2. 0.1 % Triton X-100 solution in PBS(−).

3. Blocking solution: 5 % Normal horse serum (NHS; Invitrogen) in PBS (−).

4. Primary antibodies: human-specific Nestin (1:250, Chemicon), Type III ß-tubulin (Tuj1; 1:500, Chemicon), Microtubule-associated protein 2 (MAP2; 1:200, Chemicon), human-Tau (1:250, Abcam).

5. Secondary antibodies: Goat anti-mouse IgG-conjugated Alexa 555 (1:200, Molecular Probes), goat anti-rabbit IgG-conjugated Alexa 488 (1:200, Molecular Probes).

6. DAPI (1:1000, Roche).

7. Phosphate-buffered saline solution without calcium and magnesium (PBS; Invitrogen).

8. Confocal laser-scanning microscope imaging system (LSM510, Carl Zeiss MicroImaging).

3 Methods

3.1 Preparation of Microfluidic Culture Platform

3.1.1 PDMS Mixing

1. Mix about 30 g of Sylgard 184 base (PDMS) with 3 g of curing agent at a ratio of 10:1.

2. Stir about 1–2 min to mix the PDMS well, and slowly pour it onto the silicon wafer.

3. Place in a vacuum desiccator until all bubbles are gone or place at 4 °C for 1 h.

4. After vacuuming is finished, place the silicon wafer on a 65–80 °C hot plate for 30 min (*see* **Note 3**).

3.1.2 Sample and Cover Glass Cleaning

1. Soak the corning cover glass (24 × 40 mm) in industrial ethanol for about 30 min.

2. Wash the cover glass with fresh DI water and rinse two times with distilled water.

3. Dry it in the clean bench for about 30 min.

4. Using a surgical blade, cut off around the molded PDMS.

5. The reservoirs are punched into the patterned devices using a 6 mm tissue biopsy punch to connect channels.

6. Clean the PDMS surface with the tapes and blow away with compressed nitrogen gas.

7. Check your samples are free of contamination substances and surface dust.

3.1.3 Bonding Process

1. Load samples onto a cleaned cover glass that has been treated with plasma.

2. Turn the vacuum pump and the RF power on in advance.

3. Expose the samples to plasma treatment for 1 min.

4. After plasma treatment, take the samples out and bond them by putting gentle pressure over the PDMS using forceps.

5. Check the bonding of the finished samples.

6. Once the bonding process is finished, immediately fill the devices with DI water (*see* **Note 4**).

3.1.4 PLO/FN Double-Layer Coating

1. Aspirate distilled water from reservoirs (Do not completely remove DI water in the main channel).

2. Fill the reservoirs and the entire device with Poly-L-ornithine (PLO) solution. 1 mg/ml of PLO solution is made by diluting with sterile water.

3. Place the device overnight in a 37 °C with 5 % CO_2 incubator.

4. Remove the PLO solution and wash twice with sterile water.

5. For fibronectin coating, repeat the above steps but using fibronectin solution. The coated devices filled with sterile water or culture medium can be stored at 37 °C for 2–3 days.

3.2 Human ESC Culture and Maintenance

Human ES cells (H9) were cultured on mitomycin C (MMC) treated-MEF feeder cells in hESC medium that consists of DMEM/F12 (Gibco, USA) basal medium with 4 ng/ml basic fibroblast growth factor (bFGF). Medium was changed daily and cells were transferred onto new feeder cells every 6 days using CTK solution. All cultures were maintained at 37 °C in a 5 % CO_2 incubator.

3.2.1 Preparation of Mitomycin C (MMC)-Treated MEF Cells

1. Thaw a cryovial of frozen MMC-treated MEF cells (2.5×10^5 cells/vial) in a 37 °C water bath within 1–2 min (rapid thawing) (*see* **Note 5**).

2. Take the cell suspension from the vial and transfer to a 15 ml centrifuge tube.

3. Slowly add 5 ml of MEF medium to the tube and gently mix by pipetting.

4. Centrifuge at $280 \times g$ for 3 min.

5. Remove supernatant, add 2 ml of MEF medium, and resuspend cells in the medium.

6. Plate the cells onto a 35 mm^2-gelatin-coated culture dish.

7. Culture the cells in a 37 °C incubator.

3.2.2 Maintenance and Subculture of Human ESCs

1. Aspirate MEF medium from MMC-treated MEF culture dishes in a 35 mm-culture dish and replace with human ESC medium. Incubate at 37 °C.

2. Aspirate medium from human ESC culture dish and wash once with PBS.

3. Add 0.5 ml of CTK solution and incubate at 37 °C for 5 min.

4. Aspirate the CTK solution and wash with PBS.

5. Add 2 ml of human ESC medium. Cut the human ES colonies into several pieces and gently mix by pipetting.

6. Centrifuge at $157 \times g$ for 1 min.

7. Remove the human ESC medium including the CTK solution.

8. Resuspend the human ESC colonies in the medium.

9. Transfer the human ES colonies to the MMC treated-MEF culture dish.

3.3 Neural Differentiation by Coculturing with PA6 Stromal Cells

Neuronal differentiation of H9 was induced by coculturing with PA6 stromal cells. PA6 cells were maintained according to the method described previously [10].

3.3.1 Prepare of PA6 Stromal Feeder Cells

1. Coat a 60 mm^2 dish with collagen type I (10 µg/ml) in sterile water with 0.01 M acetic acid at room temperature for 1–3 h.

2. Wash two times with PBS.

3. Wash one time with PA6 culture medium.

4. Thaw a cryovial of PA6 stromal cells (2×10^5 cells/vial) in a 37 °C water bath within 1–2 min (rapid thawing).

5. Transfer the cell suspension from the vial and to a 15 ml centrifuge tube.

6. Slowly add 5 ml of PA6 culture medium to the tube, and gently mix by pipetting.

7. Centrifuge at $280 \times g$ for 3 min.

8. Remove the supernatant, add 5 ml of PA6 culture medium, and resuspend cells in the medium.

9. Plate the cells on to a 60 mm^2 collagen-coated culture dish.

3.3.2 Neural Differentiation of Human ESCs

1. Mechanically dissect human ES (H9) cells into small colonies using a Pasteur pipette.

2. Transfer the human ES small colonies onto PA6 stromal cells in differentiation medium (DM).

3. After 4 days, change to N2 medium containing N2 supplement. Change medium every other day.

4. At 16 days, rosette-like structures appear, which are separated from PA6 cells to petri dish where they form neurospheres (NS) in N2 medium supplemented with 10 ng/ml bFGF.

5. Afterwards, the neurospheres are induced into NPCs and mature neurons by attachment on ECM (*see* **Note 6**).

6. Neurospheres (NS) are plated on the 15 µg/ml poly L-ornithine (PLO) and 1 µg/ml fibronectin (FN)-double coated device (Fig. 2).

7. Differentiate into mature neurons. Mature differentiation medium contains 20 ng/ml BDNF.

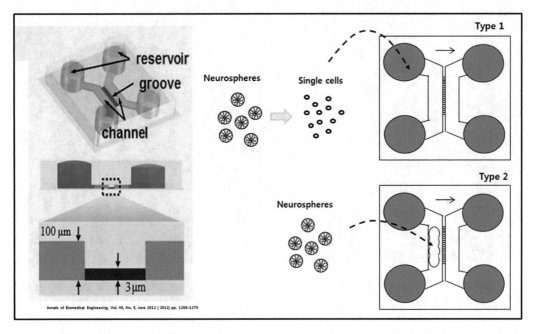

Fig. 2 Neural culture platform of the microfluidic-based chamber system. The Type 1 device was used for neural differentiation from single cells, whereas the Type 2 device was used for neurosphere differentiation into mature neurons (Reproduced with permission from ref. [11])

3.4 Monitoring of Neurite Outgrowth in Microfluidic Devices

To observe neurite outgrowth into grooves, NPCs and neurospheres (NS) were directly attached onto PLO/FN-double coated microfluidic devices for 5–7 days with the treatment of 20 ng/ml BNDF (Fig. 3). For monitoring the neuronal outgrowth, attached NS was placed inside a CO_2 incubator mini-chamber on an inverted microscope (Olympus, Korea). The medium was changed every other day, and was loaded into a line of device for 7–10 days (Fig. 4). Axons and dendrites differentiated from human ESCs on

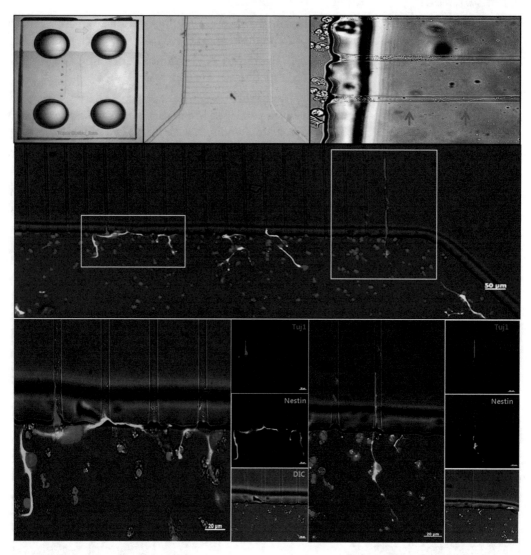

Fig. 3 Outgrowth of neural precursor cells derived from hESCs in the microfluidic device. (**A**) Overall configuration of the microfluidic device, with enlarged images (**A′** and **A″**) showing the migration of cells. (**B, C**) Immunocytochemical staining of migrating neurons. A 90° counterclockwise rotated view of (**A**). is shown. Tuj1-positive neural cells (**C**), but not Nestin-positive NPCs, entered the microchannels of the microfluidic device (**B**). Abbreviation: *NPCs* neural precursor cells (Reproduced with permission from ref. [11])

Fig. 4 Differentiation of neurospheres into mature neurons in the microfluidic device. (**A**) Overall view of a Type 2 device, with its enlarged images (**A′, A″**). (**B, C**) Enlarged images showing the presence and migration of neurons derived from neurospheres, with higher magnification (**B′, C′**). Immunocytochemical staining showing the accumulation and migration of a single axon or a bundle of axons through microchannels (**C″, D**). Type III β-tubulin (Tuj1)-positive neurons are shown in *red* (Reproduced with permission from ref. [11])

the microfluidic device were distinguished using antibodies against Tau and MAP2, respectively (Fig. 5). In this case, we observed that only Tau-positive axons are detected in the microchannels.

3.5 Immunocyto-chemistry

To analyze the marker expression of NPCs and neurons in the device groove, we performed immunocytochemistry and photographed using confocal microscopy.

1. Aspirate the medium and wash two times with Ca^{2+}, Mg^{2+}-containing phosphate-buffered saline (PBS+).

2. Fix in 4 % paraformaldehyde for 15 min and wash three times in Ca^{2+}, Mg^{2+}-containing phosphate-buffered saline (PBS+).

3. Aspirate and add blocking solution containing 5 % normal horse serum (NHS) and 0.1 % Triton X-100 solution for 30 min on a shaker.

4. Prepare the primary antibodies in PBS (no Triton X-100). The primary antibodies were used in this study are as follows: human-specific Nestin (1:250, Chemicon), Type III ß-tubulin (Tuj1; 1:500, Chemicon), Microtubule-associated protein 2 (MAP2; 1:200, Chemicon), and human-Tau (1:250, Abcam).

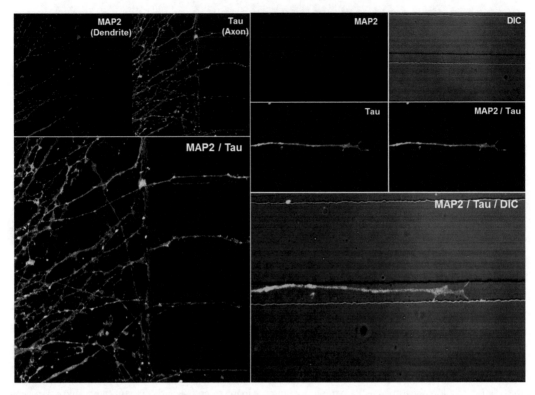

Fig. 5 Growth and migration of a bundle of axons through microchannels. Dendrites and axons are visualized by immunostaining using antibodies against MAP2 (*red*) and Tau (*green*), respectively. Note that only Tau-positive axons are detected in the microchannels

5. Aspirate the blocking solution and add the primary antibody mixture.

6. Leave overnight at 4 °C on a shaker.

7. Aspirate the primary antibodies and wash two times with PBS (−).

8. Prepare the secondary antibodies in PBS (−). Secondary antibodies used were goat anti-mouse IgG-conjugated Alexa 555 (1:200, Molecular Probes) and goat anti-rabbit IgG-conjugated Alexa 488 (1:200, Molecular Probes).

9. Aspirate the last wash and add the secondary antibody mixture.

10. Incubate at room temperature in the dark for 1 h.

11. Aspirate the secondary antibodies and add DAPI for nucleus staining (1:1000, Roche) for about 30 min.

12. Aspirate and wash two times with PBS (−).

13. Mount using Vectashield (Vector Laboratories).

14. Stained NPCs and neurons were examined and photographed using a confocal laser-scanning microscope imaging system (LSM510, Carl Zeiss MicroImaging).

4   Notes

1. Before use, FBS is heated for 30 min at 56 °C with mixing to inactivate complement. Heat-inactivated serum is especially suited for immunological work.

2. The MMC solution should be protected from light and can be stored at 4 °C for up to a week. It is cytotoxic, mutagenic, and carcinogenic, so handle with extreme care. Preparing premeasured solution in a glass bottle is helpful.

3. If a vacuum desiccator is not available, the wafer can be stored at room temperature for several hours, during which bubbles will be removed. And if a hot plate is not available, curing the PDMS at room temperature after 24 h works just as well (Do not use the dry oven).

4. The devices filled with DI water can be stored in an incubator for 2–3 days, but extra care should be taken to dry the inside channel.

5. To use feeder cells in human ESC cultures, the primary MEF cells need to be passaged 2–3 times before use to minimize the inclusion of non-fibroblast cells. Therefore, we routinely use MEF cells at P4 following MMC treatment at P3.

6. To make single cells at the NPC stage, NS was dissociated using Accutase™ (Chemicon) for 10 min in a 37 °C incubator.

Acknowledgements

This research was supported by grants from the Bio & Medical Technology Development Program of the National Research Foundation (NRF) funded by the Ministry of Science, ICT & Future Planning (2012M3A9C7050228), and the Next-Generation BioGreen 21 Program (No. PJ010002012014), the Rural Development Administration, Republic of Korea.

References

1. Reubinoff BE, Pera MF, Fong CY, Trounson A, Bongso A (2000) Embryonic stem cell lines from human blastocysts: somatic differentiation in vitro. Nat Biotechnol 18:399–404

2. Thomson JA, Itskovitz-Eldor J, Shapiro SS, Waknitz MA, Swiergiel JJ, Marshall VS, Jones JM (1998) Embryonic stem cell lines derived from human blastocysts. Science 282:1145–1147

3. Blow N (2008) Stem cells: in search of common ground. Nature 451:855–858

4. Breslauer DN, Lee PJ, Lee LP (2006) Microfluidics-based systems biology. Mol Biosyst 2:97–112

5. Millet LJ, Stewart ME, Nuzzo RG, Gillette MU (2010) Guiding neuron development with planar surface gradients of substrate cues deposited using microfluidic devices. Lab Chip 10:1525–1535

6. Millet LJ, Gillette MU (2012) New perspectives on neuronal development via microfluidic environments. Trends Neurosci 35:752–761

7. Taylor AM, Blurton-Jones M, Rhee SW, Cribbs DH, Cotman CW, Jeon NL (2005) A microfluidic culture platform for CNS axonal injury, regeneration and transport. Nat Methods 2:599–605

8. Park J, Koito H, Li J, Han A (2009) Microfluidic compartmentalized co-culture platform for CNS axon myelination research. Biomed Microdevices 11:1145–1153

9. Li GN, Liu J, Hoffman-Kim D (2008) Multimolecular gradients of permissive and inhibitory cues direct neurite outgrowth. Ann Biomed Eng 36:889–904

10. Kawasaki H, Mizuseki K, Nishikawa S, Kaneko S, Kuwana Y, Nakanishi S, Nishikawa SI, Sasai Y (2000) Induction of midbrain dopaminergic neurons from ES cells by stromal cell-derived inducing activity. Neuron 28:31–40

11. Lee N, Park JW, Kim HJ, Yeon JH, Kwon J, Ko JJ, Oh SH, Kim HS, Kim A, Han BS, Lee SC, Jeon NL, Song J (2014) Monitoring the differentiation and migration patterns of neural cells derived from human embryonic stem cells using a microfluidic culture system. Mol Cells 37:497–502

Methods in Molecular Biology (2016) 1502: E1
DOI 10.1007/978-1-4939-6478-9_5001
© Springer Science+Business Media New York 2016

Erratum to: Bioengineered Models of Solid Human Tumors for Cancer Research

Alessandro Marturano-Kruik, Aranzazu Villasante, and Gordana Vunjak-Novakovic

Erratum to: Methods in Molecular Biology (2016) 1502: 203–211
              DOI 10.1007/7651_2016_353
              © Springer Science+Business Media New York 2016

The publisher regrets that the affiliation of the author Alessandro Marturano-Kruik was incorrect. The correct affiliation is provided below.

A. Marturano-Kruik
Department of Chemistry, Materials and Chemical Engineering "G Natta", Politecnico di Milano, Milano, Italy
Department of Biomedical Engineering, Columbia University, New York, New York 10032, USA

The updated original online version of this chapter can be found at
DOI 10.1007/7651_2016_353

Methods in Molecular Biology (2016) 1502: 237–238
DOI 10.1007/978-1-4939-6478-9
© Springer Science+Business Media New York 2016

INDEX

Printed in the United States
By Bookmasters